SUPER

PIG

SUPER PIG

BY WILLIAM RUSHTON

with illustrations by Willie Rushton

Foreword by W. G. Rushton

The Queen Anne Press Division of
MACDONALD AND JANE'S
London

© William Rushton 1976

First published 1976 by the
Queen Anne Press Division of
MACDONALD AND JANE'S LTD
Paulton House, 8 Shepherdess Walk
London N 1

SBN 354 04022 7

First Reprint September 1976
Second Reprint November 1976

Design by Alan Coombes

Cover Photograph by Duffy

Printed in Great Britain by
Hazell, Watson and Viney Ltd, Aylesbury, Bucks.

CONTENTS

This book is dedicated to US

'He saith among the trumpets, Ha, Ha;
and he smelleth the battle afar off,
the thunder of the captains and the shouting'

Job **39,** 25

THE WITHOUT WHOMS

First and foremost, God bless Shirley Cooklin who master-minded or mistress-minded or indeed person-minded the research department of one – herself – and never failed to come up with such delights as the contents of a septic tank by Tuesday. Also to those singular males from whom she wheedled info. such as Peter Halliday, actor and wasp-basher, Stuart Black, Humphrey Barclay, Chris Cook, a travelling gourmet, Robin Brown, a seasoned traveller, Tony Marks, a resident landlord, Stan Katz, who was the original inspiration, G. C. Foster who revealed all about plumbing, and Norman Webster who did the same for insurance. I took on the rigorous research into drink and my thanks to Paul Frances of The Greyhound for making it a lost weekend. Mrs Beeton, some Notting Hill fuzz, Harold Macmillan unwittingly, British Rail, Qantas, the Department for the Environment, the Complaints Department of Harrods, the 3rd Royal Tank Regiment, John Wells and Cousin Tone, the Law Society for releasing me from my Articles in 1958, the Zoo, lovely Kate Ritchie for doing the typing and getting me a very nice toasted ham and cheese sandwich, and pretty, vivacious editor Rosalie Vicars-Harris, 22, who got it all together and should be getting over it all together by now. And who is Alan Coombes? Finally, thank you, wife, without whom I was at the time, who taught me everything I know?

'There comes a tide in the affairs of men when no amount of cursing will suffice –
Let us therefore remain dumb like a deaf mute who has just struck his thumb with a hammer.'   John Barrymore

FOREPLAY

What sort of a man are you? I only ask so that I know the sort of person I'm addressing. Why *are* you living alone? Why *aren't* you married? That last question is usually put by women, and intended to undermine your rugged masculinity and cause grave doubts.

There are, of course, professional celibates like lighthouse-keepers, oil riggers and popes. There are others who have been up the Amazon, and found her wanting. It's the lack of a right breast that's most off-putting, even if it does make it easier for them to draw a bow. There are those who simply hate women, others who flee in panic at the sight of them. Several who prefer their fellow-man or fancy pigeons. Some who have opted out and lie nonchalantly in their teeth under the gently rustling leaves of the marijuana tree. Widowers, divorcees, old lags, separatists, unmarried fathers, pooves, hermits and Trappists. Pigs to a man. Superpigs to a woman.

Men, simply look upon this as your life support system. I have before me as I write a picture of Shirley Conran, rather as Monty used to keep Rommel tacked to the wall of his caravan. She does not half perpetuate the myth. Or Ms.

It's a fact that 99.9 per cent of the swine I spoke to refused to believe that there is such a thing as a domestic crisis, and claim it to be an invention of the she-devil. (There may be elements of truth in this, but always remember the man who said *'In vino veritas'*, was pissed as a judge *and* lying.)

Now that *they're* liberated, and we are finally surrounded, now is the time for all Men's Libbers to stand up and be counted. One. Is that all? Oh, well, one's not bad even if he couldn't stand up. He has sniffed freedom, and in the words of the balladeer:

> *All my youth's done fail*
> *All my pep's done gone –*
> *Pick up that suitcase, man,*
> *And travel on.*

Bring up the music. Cue the sunset. Go for your life, old fellow. Sing as you go 'There's No Vinism, like Chauvinism, like no Vinism I know'.

Love Rushton

   P.S. This book contains no chauvinism whatsoever, in fact, (except for the odd references to foreigners) but I get the feeling that it will make everyone easier, if we all pretend it does. They do like to stick labels on you these days, however loosely packed.

INTRODUCTION

I would simply say that this book in no way intends to be complete. You can look up 'Grouse – How to Kick to Death in Bedroom Slippers' in the index and find not a mention. On the other hand look up 'Grouse – How to Cook in the Accepted Manner' and you won't find a word about that either. At the same time if you *have* savaged one of these poor beasts and are eager to get your teeth round it, you can always buy a cheap paperback on 'How To Cook Grouse'.

What we have tried to do is to assemble the worst 24 hours a lone man could possibly enjoy. It's the *Book of Job for non-starters in the Race of Life,* (viz. if assailed by a plague of boils – simply scrape yourself with a potsherd withal. It's full of wrinkles won in the War Against The World.)

I must admit I'm not a great Do-It-Yourselfer. There are those for instance who recommend making your own bread, blowing their own chaff and, as in Greece, crushing their own kernels. This strikes me as unnecessarily hard on those honest souls who stay up all night kneading and slicing our daily bread. The result of their labours *should* be expensive; for one thing they have to live and anyway I would pay the price of a hundred loaves not to have to slave the night away kneading and slicing.

Do-It-Yourself is fine to a point. Suppose, however, that I were to master the arts and intricacies of plumbing. That I learnt to attack the S-bend with the same vigour that Apaches used to attack the Old Bar-Q. What happens to the plumber then? He is obliged to grow a ginger beard and live by airing exploding rubbish in public. Each, cry I, to his last. And my last pulled out years ago.

FACING THE DAY

FIRST THING FIRST

It's a good idea to shout at yourself first thing. You've survived another night. You haven't died in it. As you grow older this becomes increasingly more important. So wake up, you have some cause for celebration, shout and thoroughly unsettle the system. After all, your body's been lying there complacently for six or seven hours, rumbling and gasping, winds light to variable, slight tremors in some parts, now is the time for the call to action.

'Wake up, body! Come on, brain! For God's sake, legs!' Days tend to start high-pitched anyway. It's either screaming progeny, screeching brakes, alarm clocks, milk floats whining, Tony Blackburn, cocks going off or the Vienna Boys Choir lurching home from a stag night. It's almost invariably painful. So, attack the day. Remember Jimmy Durante's old number *Start Off Each Day with a Song* – and at the time of writing he is still alive and tentatively kicking.

Here is a possible running order:

1 Turn off alarm. It's no bad idea to keep it about five yards from the bed. This does force you out of it. You can always set it again, in case of a relapse. Or use two alarms, staggered.

2 Spend a minute or two warming up the nervous system. Think of VAT or something similarly unnerving. This gets the adrenalin flowing and tightens the stomach muscles.

3 Now shout.

4 Unless, of course, you are not alone. Not being alone *can* play merry hell with a schedule. If it's an old acquaintance, she

13

will doubtless be able to fit in with your eccentricities. If she's new, it's quite a sensible idea to have jotted her name down on the pad you should always keep by the bed. (This is for 'Ideas in the Night'. You may not be able to read what you've written, but it might jog some memory.)

Actually, Mother Nature's tacky dance can be rather fun of a morning. The sexes really are equal. Equally grotesque. Equally unsavoury. Equally foul-breathed. Ah, but the heady whiff of decadence only adds to the jollity. Then, timing your exit to a nicety, promise coffee at once, and grab the first bath, while the kettle's boiling.

5 For the moment we'll take it that you are alone. So have a good roar at the morning. Have that cup of coffee. Light up the first of the day. I am aware that there are those who (*a*) don't smoke at all or (*b*) don't smoke till they've eaten or (*c*) see smoking in bed as the Original Sin, but I am just trying to ignore you. It's too early in the morning for unnecessary argy-bargy. And too early in the book.

6 Enjoy a good movement of the bowels. The simple pleasures are so often the best. All this and the *Daily Mirror* too.

7 Have a hot bath. Wash the hair. Even if it's only once through with the shampoo, and a quick blast from the shower attachment, it seems to sluice out the grey cells.

8 Clean the teeth. I am not going to advise you on how to clean your teeth. I'd recommend you get an electric toothbrush, but they make me feel giddy. I fancy, however, that they are the best. The White Teeth of the Technological Age. My teeth seem to be growing again.

9 A quick limber-up.

Nothing too drastic at this hour, you can do yourself lasting mischief. These are a few looseners I learnt from Al Murray, when we did a diet and exercise programme for the Beeb, during the course of which I am sorry to relate I put on a stone and lost the use of my left arm. I liked his style, though, he preached moderation. It *is* useful however to get the body a trifle pliable, so that when you turn suddenly to reverse the car you avoid the savage rick, or you slip a disc bending for the nasty brown envelopes with the sinister little windows. I do these simple movements religiously every morning. On my knees, gasping *Oh God, Our Help in Ages Past*.

Mild Athletic Movements for the Rudely Unfit
To get the lungs working, etc. Stand with your feet apart. About

14

To get the lungs working

1   A good loosener   2

a foot and a half apart. Say two feet apart, it doesn't look so silly then in print. No wonder we went metric. Or everyone else did. I seemed to have failed to make it.

Feet two feet apart. Hands by the side, and breathing in, stretch the arms forwards and upwards and as far back as you can, returning to original position and breathing out or coughing violently. Do this 12 times. None too viciously, either.

Do not ignore the benefits of a good cough. A really energetic hawk and spit tones up the inner man to a nicety. It sorts you out. Like kicking a sick lift. A good shake-up never hurt anyone.

A good loosener With feet in the same place as previously (you don't have to move, you see), raise the arms as illustrated – it's a position not unlike that adopted by Kung Fu perverts except that your palms face the ground in a conciliatory gesture. Then in a series of 12 graceful sweeping motions, twist from side to side, keeping your hips facing front all the time. This gets your upper body started. Once again, don't get over-excited. Relax. Something may give.

A sound stretcher Feet as before. Arms raised sideways as though demonstrating size of shark caught in strange incident in bathroom. Remaining straight-backed, not bending forwards an inch, slide your right hand slowly down the side of your thigh, aim kneewards and slowly return to first position. Do the same left-handed. Six a side will be quite enough for the moment. The joy of these exercises is that as you get looser you can do more. If you feel sudden, sharp agony, do less.

Enough is enough Or as that ancient proverb from the Isle of Skye, I think it is, puts it so admirably, and it should be painted

15

A sound
stretcher

on walls, loud and large by graffiti experts who care about the quality of life – 'Some is plenty – Enough is too much'.

Stand to attention.

Slowly bend forwards as you lift your left leg, gently lay both hands on your left calf and slowly get back to attention. Repeat with right leg. If you fall over, this in itself is no bad thing, getting up can be wonderful exercise, and set the pulse racing. Six grasps at each leg, and if you've done all four exercises you've made 48 movements, and if you're not ready for anything, you should be quite ready for something.

Another cigarette?

Breakfast?

If you really seek the Body Beautiful, go to a gymnasium at lunchtime. Simply studying the sweating executives will make you feel better.

10 I see, you're still at number **2** and unable to budge owing to a *severe hangover*. As far as my extensive researches have taken me, I have been unable to find a real cure. Had you had a pint of milk prior to setting out last night on the bat, you would feel better. If you hadn't mixed the red biddy, scotch, gin, rum, and when all else had evaporated, emptied that old, dust-covered bottle of Crème de Menthe, you would feel better. Remember when you are pouring it down yourself, God created you in his own image, which is humanoid, and did not design you as some form of fleshy cocktail-shaker.

Over-indulgence causes two things, (*a*) dehydration and (*b*) a rotten kip. The only counter to this that seems to work quite well is to down two Alka-Seltzer and about three pints of water

16

Enough is enough

THINKS: NEVER AGAIN QUITE YET

immediately before crashing out on your bed. With luck this overdose of water will wake you later. Pass it, and replace with another pint or two. You will remove some of the morning's pain.

If still bad, take more Alka-Seltzer, plenty of black coffee, read the recipe for a Prairy Oyster and forget it, and if you feel up to it, down a Fernet Branca.

Wash the glass out immediately as it leaves an indelible trail of amber crust on the side facing your face. It does seem to restore, admittedly. As a last resort, all else having failed, if at lunchtime you still feel that your brain has fossilised and it keeps rattling around in your skull and cannoning off the back of your eyeballs, have a good curry.

Apart from anything else I reckon a good Vindaloo is a cure for the common cold. Bovril – a spoonful in a cup of hot water isn't bad. If you like Bovril.

11 Have another cup of coffee.

12 Have breakfast.

COFFEE
Almost as much rubbish is talked about coffee as about wine. In both cases you know at once when you've come up against a duff cup. The urge to spit will come hurtling down from the brain in no time at all.

One of the turning points in my life was when lunching one day with a French family. I'll go anywhere for a good French lunch. At the end of five fat courses, up goes the cry *'Café?'*

'How', I wondered to myself, 'will it arrive? In some antique percolator designed by Dr Guillotine? It will undoubtedly be elegant, truly French, and the way coffee should be.' I could see the forthcoming cup or two as some sort of standard by which I could judge other cups or two for the rest of my days.

It was certainly elegantly served. Silverware abounded on a silver tray. The cups were tiny marvels. Not a crack in a saucer. White sugar and brown, cream and hot milk, and set beside a distinguished silver jug sat a family economy-size jar of Nescafé. And delicious it was. Au lait!

I have leant heavily towards the Instant ever since. A pint mug, 3 heaped teaspoons of coffee, milk and four Sweetex may not sound the height of refinement to you, but it does me all right.

(Coffee bags taste even better, but you have to wait four minutes for them to take effect, and you may not have four minutes.)

All right. So this is for you, dear:

Proper old-fashioned coffee

First and foremost your coffee must be fresh. Keep it in something airtight, and then only for a week – after that it tastes like old earth. Have the beans ground at the shop if possible. I only mention this as I am one of those who has forgotten to screw the lid of the grinder down securely prior to switching on. You immediately get some idea of how the cuirassiers felt at Waterloo when faced with the fire-power of the British Square.

If on opening the bag of coffee you find you've boobed and bought beans, and you're not blessed with a grinder, wrap the beans in a tea-towel and set to with a hammer.

The simplest method for making coffee is just to use a jug and make it as you would tea. The only difference being a heaped tablespoon of coffee per person as opposed to a heaped teaspoon of tea. Add boiling water. Stir gently. Allow to settle and stand a few minutes and then pour into cups via a strainer or a Kleenex!

The most economical method and possibly the best is to filter it, using the Melitta method with coffee ground fairly fine.

It is no mean notion if you're thinking of coffee for the end of a meal, to make it earlier and filter it into a thermos. This means you can get extravagantly drunk and not have to panic later.

Never hesitate to add a dash of spirit to the brew, even at breakfast a little whisky or brandy makes a fine 'heart starter'. A dollop of cream on top adds delight. I'm still happy with Instant + Number 6.

18

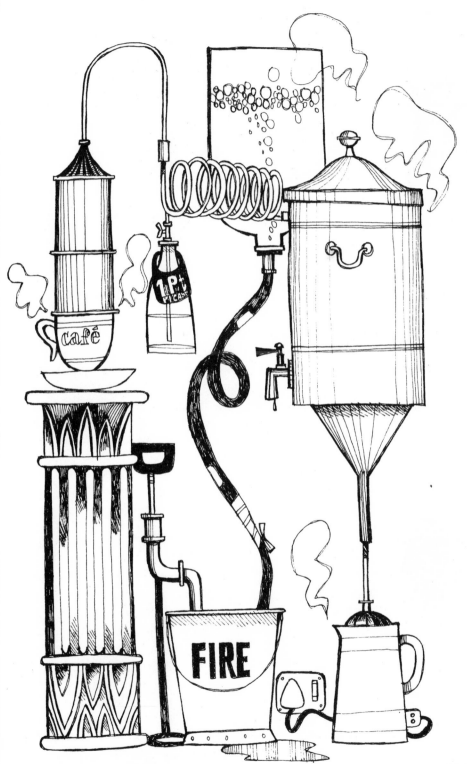

PASS THE SUGAR, G.B.S.?

BOLLOCKS, OSCAR

A Literary Breakfast

TEA

I refuse to believe that there is anything wrong with tea bags. All right, there's something wrong with *some* tea bags, but find the ones you like and you will live happily ever after.

The only mild luxury worth keeping in the cupboard is a packet of China tea. Don't forget to warm the proverbial pot before you put the tea in – 1 teaspoon per cup plus another for luck, and let the brew stand a few minutes after adding the boiling water. China tea does make a change and untouched by milk or sugar, makes a fine substitute for alcohol if you can't face it.

I will further swear that tea tastes much better without sweetening. Or perhaps that's me being smug.

BREAKFAST

I've never fancied breakfast in bed. You fill the bed with crumbs and escaped eggs, your back aches, your knees ache from balancing the tray – it's a vastly over-rated occupation. Breakfast *before* you go to bed is magnificent, and I still think it tastes better when you're fully dressed than when you're still in night-shirt or dressing-gown. It's a great meal. The brain may not be at its peak, I don't think I would have been the life and soul of a literary breakfast, but then I'm not much cop at a book bang.

In a perfect world you should always start the day with a large breakfast. It can carry you through to the evening, by which time you will have burnt up all the calories and be feeling lithe, limber and knackered. In a perfect world, of course, they do start with a large breakfast. Great steaming platters of kidneys and bacon, kedgeree and kippers borne aloft by ageing retainers or sizzling on the hot-plates under the Rubens. Any of those, of course, is within you grasp, the only problem is that you're on your tod and obliged to cook them yourself.

Nevertheless, if all you can face on a weekday is a bowl of stale roughage and half a cup of luke-warm Instant, think of these breakfast suggestions as a delightful way to start Sunday. (If you feel you owe anybody a meal, ask them round for a breakfast at noon, prior to attending your devotions at the boozer.)

Incidentally, what strange guilt grasps us in the morning and prevents us from starting the day as our forebears used, with steak and beer. I could use that on occasion. I suppose it's the sneaking suspicion that to partake of alcohol prior to opening time is but a step from the pack of green Alsatians panting at your bedhead. What a shame! A vodka and fresh orange can enliven the darkest of morns. Roll on, Sunday.

Porridge

I juddered as I wrote the word. The world's almost only grey food. Nevertheless, there are those to whom it spells bliss. Probably the same people who have cold baths and run around Hyde Park in the morning mists, frightening ducks.

To make porridge
Boil $\frac{1}{2}$ pint of water.
Throw in 2 handfuls of porridge oats (I don't know how you can do this). Add $\frac{1}{2}$ tsp salt (oh dear, oh dear). Bring to the boil and then simmer for 5 minutes.

For a creamier porridge (is there no end?)

1 cup milk	*1 cup porridge oats*
2 cups water	*$\frac{1}{2}$ tsp salt*

Golden syrup or sugar or cream can be added afterwards – I leave it to you. I'm going to frighten ducks in Hyde Park.

The Egg and You

Sitting there in the fridge is an egg. You can gauge very little about that egg from just looking at it. If you have time on your hands you can stand and marvel at it for a while. We do take them for granted.

The Brooder

Lively topics to occupy mind while peering at egg

1 Which came first?

2 How does the man who runs his car on chicken droppings get them into the petrol tank without (*a*) fearsome sounds from the chicken and (*b*) attracting the attention of the RSPCA?

Back to the point:

You have no recollection whatsoever of the history of this egg. Ask yourself these questions.

How old is it? Stick it in a pan of cold water. If it lies on its side – it is fresh. If it leans nonchalantly – it's three or four days old. If it's upright – it's 10 days old or so, but still in with a chance. If it floats – it's high time you entered politics and threw it at a Conservative. (You're allowed to be biased in a book!)

Could it be hard-boiled? It may after all be left over from one of your gourmet nights. You will enjoy this next bit. Put it on the floor and spin it like a top. If it starts wobbling all over the place it is uncooked, and you can't be doing it much good. If it spins rhythmically and evenly you have a hard-boiled egg there, old fellow, and possibly a new hobby.

I always have terrible difficulty *peeling a hard-boiled egg.* What should I do?

Dear Anxious, if after they've been boiling for 10 minutes or so, you sieze the pan and bung it under the cold tap for a while, the shell will come off with such ease.

How long should I boil an egg? Some like three minutes which is soft, some four which is a deal firmer and if it's a large egg five minutes may be necessary. In China it's a hundred years. One excellent method is to put the eggs in a pan of cold water and take them out when the water comes to the boil. You may not like it, but I do.

To avoid the egg shell cracking as you put it in the water prior to inserting the egg, put in a good dash of salt and even vinegar.

How do you poach an egg? You pinch it from a battery-hen in Lincolnshire.

Seriously? There is an American method which I have tried and I think has a certain verve. (This is if you haven't got an egg-poacher, which is a grisly thing to wash up – about eight different bits.)

Just before a pan of water comes to the boil, wait for the tiny bubbles round the side, add a dash of salt and vinegar, stir the water with the handle of your wooden spoon and produce a whirlpool effect. Launch the shelled egg off a saucer into the eye of the whirlpool and lo! You can now watch it poach.

Getting it scrambled For generous portions of scrambled eggs for two, use 3 large eggs. Break into a basin and beat well with a fork. Add salt and pepper to taste. Milk (an eggcup-full) added to the beaten eggs makes them go further but if you're wanting to impress – go mad and make it cream instead!

Melt a good sized lump of butter in a thick pan and then add your beaten eggs. Stir over a *low* heat continuously with a wooden spoon – until the eggs are cooked but still moist. The secret is to let them cook slowly and stay creamy and not form a hard unappetising lump.

Serve quickly on rounds of freshly buttered toast.

Now beat it Some of your bloody silly recipes simply say 'beat an egg' and you haven't got a bleeding egg-beater. Now what, smart-arse?

I knew this would end in unpleasantness. Either use a fork or simply break the eggs into an old jam jar and shake, baby, shake. This also saves you from having to wash up the bleeding egg-beater which always ends in unpleasantness, with egg all over the window, you and the cat.

This is not the last you will hear of eggs – the lone man's friend.

Bacon, eggs, sausage and tomato

No-one cooks bacon, eggs, sausage and tomato like you do. But isn't it strange how it always goes down easier when cooked by someone else. There are, admittedly, hotels, possibly with a chef who is continental and unable to cope with what he doubt-

less views as 'la vice Anglaise', where they fall short of perfection. I find a good dose of Worcestershire sauce a suitable antidote. It enlivens the tastebuds, and makes up for any shortcomings.

For a change also add a few slices of fried salami or black pudding. Why does hotel toast always taste of table mat?

French toast

This is a useful one if your bread is ageing. Simply dip some slices of bread into milk, then into some beaten egg and fry in butter. Sprinkle with cinnamon and brown sugar.

Bauernfrühstück au Rushton

The spelling may not be right, but, by Heavens, this is the sort of dish that has made the German what he is today. Industrious and fat. I met it years ago while holding back the Russian hordes from the NAAFI Club in Hanover. (Or Hangover, as we referred to it for clear and obvious reasons.)

Back to the Farmer's Breakfast, (for that is the meaning of that extraordinary word, which I recommend you now forget).

You will need – for 2 people:

6 or 7 eggs	*4 tomatoes*
butter	*bacon*
2 onions	*a tin of new potatoes*
a tin of corn if you like	
Peas possibly, heated separately	

24

Chop up everything, except the eggs, and start them gently frying, preferably in butter, as this makes the whole thing even more fattening and excessive.

Now take your omelette pan. Start heating it. Meanwhile, whip your eggs into a fine yellowy mess in a basin with a fork.

The omelette pan should now be hot enough to throw in a good lump of butter. At once, it will start to sizzle and melt. By dextrous manipulation, make sure the butter covers all the pan, and empty in the beaten eggs.

Now it starts to get exciting.

Almost at once, your omelette begins to grow before your very eyes. Keep it moving. By lifting the edges of the cooked portion and allowing the uncooked goo to slide under the outer skin and meet the sizzling pan, its size increases rapidly.

Keep it rotating until quite soon you have an omelette staring you in the face. The inside should still be pleasantly moist. Now pour in the filling, that heady brew of fried bacon, onion, tomato, spud and anything else you cared to add, undaunted fool. Fill a complete semi-circle of the omelette, and then with your implement fold the lid over. Now serve.

Not such exciting omelettes, but none the worse for that:
Grated cheese omelette
Fried mushroom omelette
Cold chicken omelette
Chopped, drained, tinned pimentos omelette
Anything you bleeding like omelette
If in doubt, practise 2-egg omelettes on your own. You only learn from your mistakes. You don't *have* to eat them.

Kedgeree

And this is the answer for your leftover rice. Melt some butter in a good, heavy saucepan and mix in the rice and flakes of cooked, smoked haddock. (They sell them frozen so you can use half of one now from the freezer, and depending on whether you find you are the master of kedgeree or not, pop the other half back, wrapped in tinfoil or give it to some roaming Tom Moggy.)

Add plenty of salt and pepper.

Heat away until very hot, stirring madly, to prevent those ugly moments de-ricing a saucepan. At the last moment, add 2 well-chopped, hard-boiled eggs and the top of a pint of milk.

If you don't feel up to it now, put it in the fridge and have it tomorrow. If you have some croissants in the freezer (good notion) heat them in the oven and go continental.

Tinned cod's roes on toast

There's easy. All you do is warm up the cod's roes and slap them on buttered toast with a squeeze of lemon.

Kippers

Here's traditional. The least offensive way of doing kippers is buttered, sprinkled with lemon juice and put under the grill. Give the scaly, fleshy side 2 minutes or so and the 'bone' side 3 minutes. I say 'least offensive' because you can just fry them in butter, but your kipper is a game little fellow and his last throw will be to make certain you never forget him. It's not a pretty aroma, but it has tremendous staying power. Apart from which, whatever you fry next will also smack of kipper.

Serve with plenty of bread and butter.

Nepalese scrambled eggs

You weren't expecting this either.

Put 2 tablespoons of butter in a thick pan, with a dash of chilli powder, and start frying a well-chopped onion until it's pleasantly soft, but hasn't started browning.

Meanwhile, whisk up 4 eggs in a bowl, and add about $\frac{1}{2}$ teaspoon of ground ginger, a little finely-chopped parsley and salt. Then toss the mix into the frying onion and proceed as per your usual scrambled egg, page 23.

It makes a change and scares hell out of a hangover.

Fried matzos

Here's different. Most delicatessens will have packets of

I WAS SITTING DOWN WRITING THIS BOOK TILL I DISCOVERED NEPALESE SCRAMBLED EGGS

matzos. Simply break them up into small pieces and drop into a bowl of water to soften them. Drain off the water when they're soft, any excess you can squeeze out with your fingers. Melt butter in a frying pan, add matzos and a couple of beaten up eggs. Add plenty of salt and pepper and fry until they're crisp. Turn frequently. Not you, revolving idiot, the fried matzos.

CLEANING UP IS SO VERY HARD TO DO
– especially after the night before

Invest in those large *black plastic bags* – better and more hygienic than any dustbin. Some local councils will even issue them free. Every morning drift slowly round house, flat or room dragging it behind you like a Neanderthal fiancèe. It is designed after all to take over one facet of woman in your life. Put in it yesterday's newspapers, all useless envelopes, letters you needn't answer, bills that are still in black, begging letters from the Inland Revenue, empty bottles, broken glasses (first wrapped in plenty of newspaper), socks with holes in, failed underpants, all kitchen unpleasantness, contents of ash trays, (give them a wipe with inside of bag), messages to yourself that you have answered, old bread, surplus milk bottles that will otherwise take over your life, surplus wire coat-hangers likewise.

Having done the circuit, put the bag, if it's not full, in one of those swivel-top plastic dustbins that should be in every kitchen (they're twice the size of those kitchen pedal bins, and five times more use). If the bag is full, secure the top, and put out instead of a dustbin and cheer your local refuse operative's day.

In praise of the hoover

Get one of those machines with a good, long extension pipe, countless attachments and a handy, mobile body that will follow you anywhere.

Make sure the electric lead is so long that you only have to plug it in once.

You can hoover anything. Not just floors, carpets, but window sills, curtains, blinds, mantlepieces, bookshelves, books, the insides of drawers and cupboards, lint-infested clothing, your trousers after a distressing fall of plaster, run it over your shoes with the brush attachment, treat it as a friend. I have seen Masters of the Hoover catch wasps down it and dozy flies.

Keep the disposal bags, they're expensive, and you've never got one when you need one. Simply tip bag into your large black plastic friend.

29

BED MAKING

With a duvet, this is simplicity itself.

Without a duvet, leave the bed exactly as it was when you so reluctantly left it. After all, wasn't that the most comfortable it's ever been? Ever?

If you are *not* the proud possessor of a duvet – get fitted sheets – these are elasticated at the proper corners and don't slowly clamber up the bed towards you like a Ku Klux Klansman on the razzle.

Try to avoid nylon. It could be me, on the other hand it could be you too, but I find you can develop severe elbow-burn, particularly when cornering.

THE TRIVIAL ROUND, THE COMMON TASK
CAN DRIVE YOU TO THE FRIENDLY FLASK

WASHING

What does tend to mount up in vast, unseemly heaps is the washing. It's quite a sound notion to keep another of those flip-top bins in the bedroom for instant disposal of shirts, socks, underpants, etc. But, alas, they cannot live there forever. Paper underpants enjoyed a brief vogue, but about three hours after their maiden flight they hung about me like a grass skirt in a fourth-rate panto. Blasted to shreds by the rigours of life.

Rule 1 Strike up a friendly relationship with the lady at your local launderette

Given this you can pursue this simple plan. In the morning, drop off kid at school, and load of bag-wash at launderette.

WRONG

When picking up kid, pick up load of bag-wash, which that sweet and exotic lady has done for you. That is life made simple. Perhaps you're obliged to take the harder path. Staggering Christian-like under your enormous burden you journey through the Despond of Slough or Rotherham or wherever to the launderette . . . perk up! The trumpets may yet sound on the other side. They are sociable places. There are always people there to guide and assist.

Rule 2 What not *to take with you*
Non-iron shirts
Drip-dry trousers
Silk shirts or scarves or handkerchiefs, you Flash Harry, you.
Anything woollen
Anything whose label proclaims loudly that it needs dry-cleaning.

Rule 3 What to take with you

Hot wash	*Warm wash*
Towels	Nylon socks
Tea-towels	Coloured nylon underpants
Handkerchiefs	T-shirts
White cotton underpants	Non-iron shirts you'd rather
Cotton shirts	iron the collar of, than wash
Pyjamas	Acrylic sweaters
Sheets (if you must)	

Some wise words about washing woollen sweaters

Read the labels on your sweaters. (One night when you're alone in a hotel room near Aberdeen, you've finished the Gideon

31

Bible, and are dying for something else to dip into before retiring.)

The *synthetic* ones – the Bricrimponylocourtelan, etc., will have washing instructions attached, anyway they can go to the launderette, the warm wash. They're man-made and can look after themselves to some extent.

Woollen sweaters are another bottle of fish. Do not, for instance, think that because the label says 'This sweater is Machine-Washable' it can go to the launderette. Not so, it means buying your own washing machine, with all the appropriate programmes.

Woollens, cashmeres and mohairs *must be washed by hand.*

What about dry-cleaning woollens?
(That's a very good question, and I must admit that like you, I would sooner rollerskate through Dante's Inferno, than stoop for a second over the sink.)

a It's expensive.

b It tends to shorten their life expectancy and they don't smell so good.

c Not only won't dry-cleaning remove all the stains, it can actually set them permanently.

d I agree. It still saves a lot of bother, doesn't it?

How to wash a woollen sweater by hand
This is addressed to those who care, and I think none the worse of you for that. Indeed I envy you.

Rule 1 Never let them get too dirty. It kills them off faster.

Rule 2 Wash white sweaters separately from coloured ones.

Rule 3 Don't have any white sweaters. You're asking for trouble. Very well, you play a lot of cricket. Point taken.

Rule 4 Check your coloured sweaters to see how far they run. You can see if they're runners by dipping an unimportant part of the sweater into warm, soapy water. If any colour emerges, never wash it with the other sweaters.

A right old pooftah you will look going in at 8th wicket down in a pink woollie.

Rule 5 If, and you probably are, (always put off till tomorrow, what's a pain in the arse today – Old Saw*) washing all your woollen sweaters at once – start with the palest.

Rule 6 Don't soak them. If they are mildly disgusting you can give them an hour, but no longer, in cool soapy water.

*Further, Old Saw, you're wrong.

32

♪ AH, BIDETS ARE HERE AGAIN!

Are you stooping comfortably? Then we shall begin.

1 You'll need two receptacles. If you have a double-sink you're laughing. If not it's basin and bath, or basin and a plastic bowl or bucket. A friend of mine swears by the bidet. I quite often find him there crouched and blaspheming.

2 Fill basin with lukewarm water and enough soap flakes to create a rich foam. *Let the soap dissolve* before immersing sweater. Hot water, incidentally, encourages shrinking.

3 Be gentle and kind with the sweater. Don't strangle it or maul it. Has a sheep ever done you any harm? If there are any particularly bad patches, just work a few soap flakes in with your finger.

4 Rinse immediately in the bath, bucket or bidet. Lots of lukewarm water, and give them a good swirl. Think Dervish. Change the water at least three times. You'll know when the last soap's vanished, when the water is clear.

5 Lay the sweaters, one at a time, in a towel and roll them so that the surplus water is gently squeezed out.

Or, and this is a lifesaver, either invest in a spin drier, borrow one from a neighbour, or use the one at the launderette – but put sweaters in the spin-drier for 15 seconds.

6 Drying. Don't put woollen sweaters on a radiator, or near any direct heat or in any occasional sunlight. Lay them flat if possible. Bedeck the bathroom floor with newspaper and lay them on that, or if you haven't done the crossword, on a towel.

Silk

You may have something silken, something of Granny's, a

33

scarf, a shirt. Wash it in the same manner as you would wool. Lukewarm water, gentle squeezing, no wringing, but thorough rinsing and roll it up in a towel to get the surplus water off.

Don't spin dry.

Dry silk away from heat, particularly if it's white, or you'll get strange brown scorch marks.

Iron whatever it is before it gets completely dry, with a warm iron pressed to the reverse of the thing.

Jeans

Most of these are pre-shrunk nowadays – the label will doubtless tell you if they're not. Gone are the days when you sat trousered in the bath, or strode purposefully into the briny, remembering or forgetting to empty the pockets.

If you needs must wash them by hand, soak them first for a bit if very soiled, then as for wool, although you can be slightly more brutal with them. Use a nail-brush on the more appalling patches.

Smooth all the creases out when you hang them up to dry – you shouldn't have to iron them. I mean, what is the point of jeans if you have to iron them?

When they become totally disreputable: (a) sell them for a fortune in the King's Road or (b) if the knees have gone, cut off the legs above the knee and you have sprauncy shorts for your 14 days in Nirvana.

JAZZING UP YOUR JADED UNDERWEAR

You can choose from 15 glorious shades and dye all your greying underwear the same exotic colour – and be the envy of all who burst into your boudoir. You can buy already mixed instant dyes to use in a washing machine or with very hot water in a sink or bath. Follow the instructions with the dye very carefully, particularly in relation to which dyes are suitable for synthetic fabrics, cotton, nylon, etc. If you haven't got a washing machine and don't fancy blocking up the landlady's sink or bath, take everything to be dyed to the launderette and chat up the assistant, otherwise she may have a small attack when she sees the purple people-eater eyeing her fish-like from the machine. To be sure of removing all trace of dye from the machine thus rendering the next wash an unseemly shade of turquoise, run the machine again empty, save for a cupful of bleach. This purges it.

While at it why not *dye your dirty-white sheets* a deep

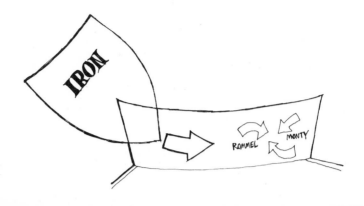

masculine blue or dark brown – much smarter than dirty white, and hides a multitude of sins. Check carefully with the shop assistant, or read the instructions on the dye and see that you purchased the correct one. You can get one made by Dylon which washes and dyes at the same time. Dylon Super White does amazing things to garments like cricket flannels or surplices which need a crash treatment for instant whiteness.

HOW TO IRON A SHIRT

If you must. (When in doubt simply iron the collar and wear a waistcoat or tasteful jumper.) If you insist – (unless you have a steam iron that spurts) – the shirt should be damp.

1 Iron the inside of the cuffs first. Apart from all else it gets your eye in. Now iron the outside of the cuffs.

2 Iron the sleeves.

3 Iron the back of the collar first and then the side which shows. Start at the points and move the resultant wrinkles in towards the middle.

4 Iron the inside of the back of the shirt. Prostrate the shirt on the ironing board as you might Mrs Bradley, unbuttoned and arms akimbo.

5 Bring up the sides as though restoring Mrs Bradley to her previous modesty, but do not rebutton until you have ironed the front, concentrating on *a* the collar-bone area and shoulders and *b* round the buttons and button-holes.

6 Button up and add finishing touches.

7 Fold as illustrated (over), *or* stick it on one of your 783 wire coat-hangers. (I have a theory that wire coat-hangers breed.)

36

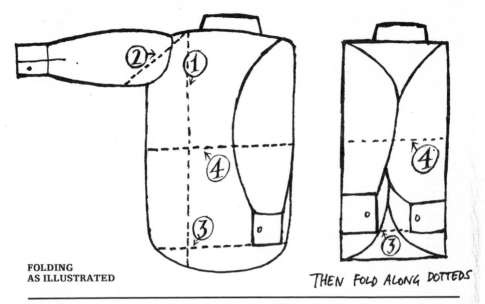

THEN FOLD ALONG DOTTEDS

If ever forced into stiff or semi-stiff collar and you find it so starched you can't get the studs through – give the relevant bit of collar a lick and a suck and it's yours.

This also applies on the rare occasion you wear a wing collar – lick the pointed bits thus blunting them and you make it through the soirée without severing your jugular.

SEWING ON A BUTTON

Always keep two needles ready – threaded when the hand was steady. One should be threaded with white cotton which should handle the shirts and the other slightly more robust, with black or brown thread for jackets. Having threaded the needle, double the cotton back, so you have two equal lengths of about a foot, and then knot.

(*Cheering thought:* There are quite a number of buttons that are quite unnecessary and can stay off. Don't let them bully you.)

Simply comparing the unsewn-on button with the others that *are* sewn on is quite a satisfactory guide. Just remember: *(a)* Not to sew it on too tight or it will never button up again. *(b)* Not to use glue. *(c)* Not to bleed too profusely over the garment.

Getting started

Push the needle through material from the inside or reverse of the material, as far as it will go, the knot will stop it passing straight through, and bang it through a hole at the back of the button and back down through another hole, back through the material and already you're in control, the button sitting once more in its old home. The knot underneath is a good target to aim at when pushing needle through button.

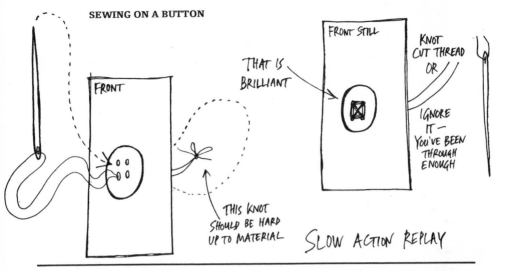

SEWING ON A BUTTON

FRONT

THAT IS BRILLIANT

FRONT STILL

KNOT CUT THREAD OR

IGNORE IT — YOU'VE BEEN THROUGH ENOUGH

THIS KNOT SHOULD BE HARD UP TO MATERIAL

SLOW ACTION REPLAY

On the return journey, go back through the hole opposite or diagonally opposite. Some buttons only have two holes, which cuts down your options by half and prevents unnecessary fretting.

(Quick research through wardrobe reveals that the old four-holed button seems to be giving way to the two-holed button. This could just be me. I've never looked before.)

When to stop

1 Compare with the other buttons and see how much thread they've used. Use a bit more, because if they had, yours might not have come off.

2 Give button a tug and rely on your male instinct.

Stopping

1 Go back through a hole but not through the material, and then wind the thread around the stem, so to speak, six or seven times. Then continue on through the material.

2 To finish off – either do some unfancy stitching at the back of material *or* cut thread and knot the two ends together.

3 Do not leave needle dangling.

Must I really go through all this?

Well, there used to be a wonderful invention called a *Buttoneer* – a sort of hypodermic which injects small or large plastic grips through button and material – be it tweed or leather. It is now only in the shops at Christmas but you can get it direct from Ronco Teleproducts, 111 Mortlake Road, Kew, Richmond, Surrey.

39

CLEANING THE CARPET

If you can't afford or tolerate an annual doing-over by a team of professional carpet-cleaners, which is only really worth it if you are wall-to-wall all over and have a number of walls then buy a large can of carpet cleaner and follow the instructions on the can.

The best I've found are those foamy ones that you squirt on, spread lightly with a clean plastic broom (never bend unnecessarily), and hoover up once it's dried. The advertisements suggest that small bodies in the foam burrow about in your carpet, gourmandising away on old ash and sundry muck, and are content, thus sated, to be sped into limbo by way of the hoover. Who are we to disagree?

One small thought. No-one is to know your carpet's actual foulness unless you have been foolish enough to clean a small section in the middle. It's like drawing a small heart. With a carpet, you clean a small area in the middle and an hour later you're still richocheting from wall to wall. So sit by the newly clean area, assay *The Times* crossword, and spill ash slowly over the offending patch, grinding in with circular movements of the foot until it matches up with the rest.

STAINS

The essence of stain-removing is *speed*. In the main, soap and water clean off most of most.

It is possible of course to turn them to advantage, namely lipstick on the collar. Get a lady to kiss your shirt all over. All your friends will want one.

40

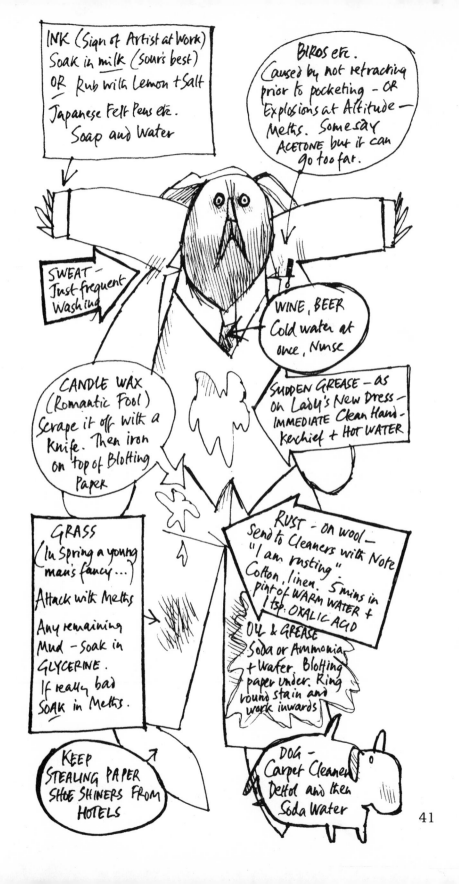

41

HOW THE KITCHEN WAS WON

First of all, surround yourself with as many of the following items as you can. (Literally surround yourself if possible, then all you need to do is rotate gently at about 5 r.p.m. Here's a layout suitable for the single man in a perfect world. It's not meant to be easy.)

Rolls of paper towels These are life-savers, and can be used all over the place. Floor, kitchen-tops, nose-blowing, child-dusting. There was a butcher during the lavatory paper crisis of 1972 who made a fortune chopping them in half with his cleaver and flogging them as a rough and ready alternative.

Rolls of tinfoil Again a valuable companion. If you're grilling sausages, say, under the griller, rather than have a tray full of charred fatty remnants, spend a happy moment manufacturing a tinfoil tray large enough to house the sausages, bent up all the way round with sturdy corners, and save yourself a deal of washing up. You can also wrap almost anything in it and keep it in the freezer.

A large plastic bin I've mentioned this, with a black plastic bag inside and a flip-top or swivel lid. I only mention it again as a bowel-chilling memory just flashed through. One day I put the laundry out in a black plastic bag, and the dustmen came a day early.

A blender or liquidizer Excellent for fast soups, milk shakes, gin fizzes, and for making a good deal of noise if you want to sound busy. Usually they come with a book of recipes, and quite good they are.

A pressure cooker My cousin Tone suggested this and he used to be highly domesticated. It cuts time by half, he says, and once again you get a book of recipes with it.

A dish-washing machine Well worth saving up for, or chancing a small bank job. You can operate with fewer plates and less cutlery and only remove them when you need them. What a boon.
Otherwise a *double sink*. Although this just gives you twice the space to forget everything in.

42

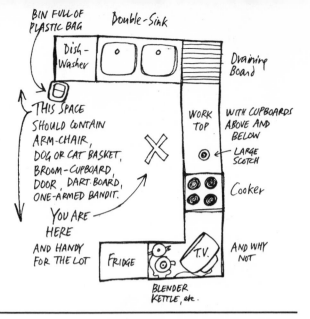

BIN FULL OF PLASTIC BAG

Double-Sink

Dish-Washer

Draining Board

THIS SPACE SHOULD CONTAIN ARM-CHAIR, DOG OR CAT BASKET, BROOM-CUPBOARD, DOOR, DART-BOARD, ONE-ARMED BANDIT.

WORK TOP

WITH CUPBOARDS ABOVE AND BELOW

LARGE SCOTCH

YOU ARE HERE

Cooker

AND HANDY FOR THE LOT

FRIDGE

T.V.

AND WHY NOT

BLENDER KETTLE, etc.

If you only have a single sink – get a large plastic bowl and proceed to wash up as described below.

Therefore a good washing-up implement With a long handle. Either a sponge or bristle, but without which life is hell.

Brillo pads and *J-Cloths*.

An electric kettle that turns itself off – *or a whistling kettle* – but you may be out of earshot.

A tool box

I'm not recommending anything ambitious, unless you're by Black out of Decker, but simply a cardboard box under the sink containing: Light bulbs of all appropriate shapes and denominations. Plugs and adapters, likewise. A pair of pliers; fuse wire; a hammer; picture hooks; picture wire. Screwdrivers, to handle plugs and fuses. Nails and screws, tacks, drawing pins. Insulating tape. One or two good long extension wires, fitted with plug and the male/female bit ready for anything. A torch. A gimlet, an aid to screwing. (So, of course, is the other gimlet, gin, lime and a dash of iced water.)

Where I stop is one after rawlplugs and the like. If you can handle things after that you're in another ball-park, dearie, and good luck to you.

Washing-up

The secret of washing-up is to have a sinkfull of hot water and a washing-up liquid that is kind to your rubber gloves, but beforehand to soak everything as soon as possible after its finished with. The fact that three days later this has congealed into an evil-smelling, septic tank packed with pans, plates, bacon

43

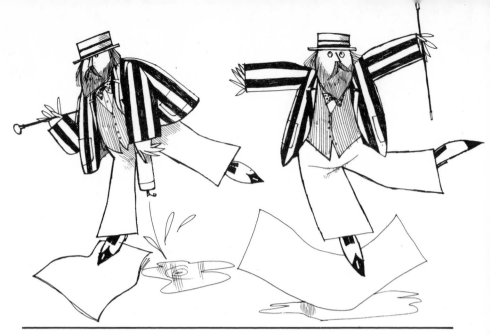

rind, and floating fag-ends must not depress you. Pour yourself a large one and attack it. It still comes off easier. If you have had a glass of milk (pre-Tiles) wash it out at once in *cold* water or you will live to regret it. I have spoken.

Cleaning the kitchen floor

1 Hoover kitchen all over first – floor, tops, tables, etc. Give the working areas a wipe with Vim and a J-Cloth.

2 Squirt washing-up liquid here and there on floor and, having dampened a tea-towel due for launderette, if you haven't got a special floor cloth, work it around floor with feet in a rhythmic dancing movement.

3 Dry off with an old towel due for a wash, using same intricate Fred Astaire-like shuffle.

4 For the black heel-marks left by previous dancing without towel, use lighter fuel.

5 Unpleasant lumps of dried-up gunge can be removed easily by a scrubbing brush, Vim and hot water or taking off the attachment of the hoover and scraping up with the tube-end.

Cleaning windows

Window cleaners being a thing of the past, since they hung up their ladders, bought cradles and devoted the rest of their lives to cleaning 160-floor crystal palaces, the only solution is, on occasion, do-it-yourself. Not having any sort of head for heights, I have to operate from indoors and still bend towards the system I learnt during my two years before the Colours – Brasso and newspaper. There are of course other methods like chamois and aerosols – *chacun à son* perversion.

Cleaning a stainless steel sink

Butter a few thicknesses of paper towel. This is not a recipe, this is how to clean the stains off a stainless steel sink. I worry sometimes.

YOUR FRIEND THE FRIDGE

'I haven't had food poisoning now for eight or nine days', confessed one of our lone interviewees, smiling the while. 'Actually, the best cure for food poisoning is carbon. Your burnt breakfast toast is the finest cure there is – no need at all to waste a lot of money on expensive carbon pills from the chemist.'

He's lived long and well, but then so has the old gentleman down the road who's smoked 80 Capstan Full Strength, man and boy, and convinced you that there's no justice.

Some simple rules for the fridge and the things that lurk in it:

Meat

Basically, the larger the lump thereof, the longer it will keep. Much depends, of course, on how fresh it was when you bought it, but in the main you can trust: A large joint for five to six days; a whole bird for three days; bits of bird and chops and steaks for two days. Minced meat and offal for one day only.

Meat should be put away unwrapped on a plate – especially mince, which can look like an exploded mind in no time. You've got to let the air circulate.

Take the giblets out of a chicken before insertion in the fridge. If you're squeamish, put on a rubber glove and pretend you're a best-selling vet.

Fish

I'd use it at once personally but otherwise put it in a container with the lid on, otherwise its basic fishiness will spread. Use within 24 hours.

Fruit

Oranges, lemons, grapefruit, pineapple, melon, grapes can all be stored in the fridge, unwrapped. Apples are happiest in a cool, dry place, but are alright in the fridge.

Raspberries and strawberries and other softies are best eaten at once, and quite right too. Pick well over, removing any bruised, bad or doubtful fruit. Take off leaves, etc. If storing, spread well out in fridge. Wash, as little as possible and just before eating. Once washed, soft fruit tends to go mushy.

Never be-fridge bananas; they go black, and startle the neighbours.

Vegetables
The main point to remember is never to put in the fridge anything unwashed that may be harbouring insects or wildlife.

Potatoes should only be put in if cleaned and peeled, and only then for a day and covered with cold water, else they too go black, and you know what the neighbours are thinking.

Tomatoes, peppers and cucumbers last longer in plastic bags or a lidded box in the fridge.

Leafy or root vegetables Remove the unusable bits, wash in cold water and dry before putting in the fridge.

Lettuce should be washed very thoroughly. Heaven knows what isn't resident. Shake and dry well and store in the salad drawer or in a plastic bag.

The raising of lettuce – a miracle. You have a dead lettuce. Dip it in hot water. Dip it in iced water with a dash of vinegar. Stuff in the fridge. Hello, Lettuce, welcome back.

Left-over cooked food, such as meat
These dry out at pace unless encapsulated in covered containers or in a dish covered with those brilliant self-sealing, plasticy, filmy, stretchy materials.

Half used tins of stuff
Despite rumours to the contrary, you can put these straight into the fridge. Only the acidy ones like grapefruit or tomato pick up the tinny taste. If in doubt just decant into a bowl.

Frozen foods
As the name suggests should be kept in the freezer department. Simply follow the directions on the packet, and don't re-freeze after once thawed. If vegetables have defrosted in your shopping bag before you get them home you should cook and eat them straight away or eat later cold as part of a salad.

Bread
This stays fresh if wrapped in foil in the fridge or freezer. If it's sliced it toasts while still frozen, a fact occasionally worth remembering, particularly if you need to burn some quickly (see food poisoning page 45). If you don't want to toast it and it's frozen, hold a hot iron half an inch above the slice. Repeat on the other side. Amaze your friends by ironing bread.

The Dying Crafts : 27

Ironing Bread

Drinks – a careless note about lager

Beware of putting bottles in the freezing compartment to chill. They tend to explode, if forgotten.

Caring for the friendly fridge

1 Keep the book of instructions, and in particular the bit about defrosting. *Defrosting* should be done when ice builds up to about $\frac{1}{4}$-inch thick, or about once a fortnight, depending on how often you leave the door open. As you have probably already lost your instructions this is how to defrost.

Each fridge is different and some defrost automatically, but most have a 'defrost' button or you turn the regulator to 'defrost' and go quietly away. It's a good idea to let this all happen over night and when your food stock is low particularly if your ice-box is so frosted over it looks like an igloo. In that case you will need some kind of receptacle in addition to the drip tray below it, to collect the melted ice as it crashes to the floor – otherwise you will have a flooded kitchen. If you defrost regularly it is relatively painless and your friendly fridge will function much better.

When it's all over wipe out and wash the inside of the fridge shelves and containers with warm water to which you have added bicarbonate of soda (1 tsp to 1 quart of warm water). Use nothing else. No scouring powders or detergents. Rinse well with clean water. Fridges need regular cleaning to be effective as a hygenic means of storing food. Mop up spills as soon as they happen or the dreaded lurgies will spread and you will end up with food poisoning like our belated friend.

2 *Don't* overcrowd your friendly fridge.

3 *Don't* put hot food into it.

4 *Don't,* well try not to, leave the door open – awful things can happen . . .

A CLEANING LADY

Weighing in at about 70p to £1 an hour, depending on where you live, what a boon and a blessing they can be to single men.

How to come by a cleaning lady?

a Ask any friends if they've got one already who might be eager for a further quid or two per week.

b Put an ad in your newsagent's window. This may also bring you strange offers from deviants, but it seems to work.

c Thumb through your local press. Not so likely, but possible, if you're after someone more ambitious – cook, child-minder, hooverer/gardener, etc.

When conducting embarrassing interview ask her if she cleans ovens. There are many who don't apparently, either because of bad backs or for religious reasons. If she says she does clean ovens, at least you've opened up an area of conversation out of which may emerge some of her own special quirks. I knew one whose sole talent was the cleaning and ironing of J-Cloths. Everything else was revolting.

There seem to be two schools of thought about cleaning ladies, one represented by a bachelor I spoke to who said, 'Every Wednesday I get into a total frenzy when Winnie's about to come, cleaning up and making it tidy enough for her to come and clean up'; and the other school, who rush about scattering rubbish and use up any number of plates and pans so that she's got something to do.

General rule – Go out, leaving a note of what you'd like done and her money. The most useful tasks are cleaning kitchen and bathroom, and windows if possible and ironing and sewing, plus washing up.

The main problem I discovered in men's minds was getting rid of them, (cleaning ladies), usually on the grounds of incompetence. No one seemed capable of looking the creature in the eye and saying, 'Out of my life you rat-bag, and never whiten my doorstep again'. This is not surprising, as we are a sweet and generous sex.

Tell her either that you are leaving almost immediately for foreign parts or that sudden financial ruin means a cutback and she's it.

TO BE BRUTALLY FRANK, SIR, I DON'T DO OVENS. I DON'T IRON OR SEW. OR CLEAN WINDOWS. OR CLEAN. OR ENTER BATHROOMS. I TEND TOWARDS MAKING A CUP OF TEA AND DISCUSSING MY LEG.

WHAT IS A 'HOVEN', DARLEENK?

AU PAIRS

More trouble than they're worth. You'll resent her if she's ugly and industrious and if she's pretty she'll lie about all day prettying herself further, cause your knees to sweat, and probably end up 'up the Community', either by (*a*) the postman or (*b*) and infinitely more embarrassing, yourself.

Apart from which, you'll spend a deal of time translating and interpreting and clearing up after her. And baby-sitting for her. And taking telephone messages for her. And wrestling with the Home Office. And turning the telly volume up so that you can't hear the gasping and frenzied smack of fleshy matter from her room. And – oh, just get yourself an 'old dear'. And British to boot.

THIS IS *NOT* YOUR LIFE

'Don't Do-It-Yourself' is my battle-cry, you've probably heard it before. You have no redress against anybody if it goes wrong. There are odd things, however, you can have a crack at round the home.

ELECTRICITY

Changing electric plugs

If you can't do this with easy grace, your local branch or shop of the Electricity Board will give you a simple map. Quite often, if you feel the plug is at fault, or even that the hoover or telly have finally died, it's simply a case of the tiny fuse inside the plug having known better days. Keep a supply of them handy.

WHAT IS THIS?

(a) A JEW'S HARP?

(b) A FUSE AS IT SHOULD BE?

(c) ANOTHER JEW'S HARP?

Mine seem to blow quite often. You can't miss the little cylinder once inside the plug, and they're a simple business to replace.

Mending a fuse

Go to your fuse-box and turn off the mains switch. Remember a torch if it's dark. Pull the fuses out one by one till you find the blackened one with a broken wire. Unscrew the little screws at either end. Remove the devastated wire and insert a new stretch of the same size. It should look like a one-string Gentile Harp when done. Plug it back. Fire the mains switch. And behold again.

PLUMBING

Since the Flood, Man has been rightly wary where water is concerned. In the average English home we have some control over the four basics, Fire, Earth, Air and Water. *Fire* we can guard against to some extent and insure against comprehensively. (Though if you do live in a fourth floor flat and there's only one staircase, do invest in one of those simple rope fire escapes, apart from which it's quite fun to test.)

Earth, as long as your foundations are solid, your sewage system operates and you have the necessary damp courses this shouldn't bother you unduly. There are of course moments when they start mining under you, or building the New Victoria Line, and you wake up in the basement.

Air is a matter of taste. I've come to actually enjoy the flavour of London air, and when living in the country, used to have to go to the garage and run the moped for a while so that I could breathe in some health-giving exhaust – start the system up –

52

get the fumes down me – the mixture's a bit thin in the country. The air a bit fresh.

Water now is something else. You take it so for granted and yet some months ago I realised, as did the *Marie Celeste* in her day, that we are ever at its mercy.

I was cleaning my teeth one morning and realised that I was singeing my gums with boiling water. Closer inspection revealed that I was using the cold tap, but there was no doubt about it, it was exuding boiling water. Because my brain at that time of day is sharp as a mountain ant's I tried the hot water tap and it was freezing. Same with the bath. Same in the kitchen.

I went to the cold water tank, a large black plastic bucket over the kitchen door, it was bubbling like a witch's cauldron. The electrically-operated hot water tank stood shivering with cold, lagged as it was. I turned all the taps on (which I later heard was the correct thing to do) and rang the Water Board, who clearly hadn't been taken over yet, as they disclaimed any responsibility for bubbling cold water and suggested the Electricity Board who were round like a shot and fixed the thermostat on the boiler.

Suffice to say as a result of this trauma, I chatted up a plumber (at a cocktail party at the House of Lords, incidentally, but that's something else entirely), and he mooted the following – (clearly biased, he is still a disciple of my basic belief in Don't-Do-It-Yourself):

Insurance
Make sure you're covered against water damage. All things, however good, wear out in time. Who said 'A thing of beauty is a joy for a fortnight'? I did. Thank you.

Maintenance
Find a good plumber, or plumbing company, the old individuals with lads appear to have retreated more and more under the wings of companies. Ask for a quote for a yearly check on the system, the hot and cold water supplies, the hot water tank (the thermostat if it has one, the lagging) and the *drains*. What goes on down your drains, sir, is nobody's business. Of course, it's *his* business. So he probes them annually – you won't get a build-up of calcium, caused by nature's own drain-fodder, that narrows the exit and causes disaster.

Central heating
Your radiators and pipes, assures the drain-brain, should be

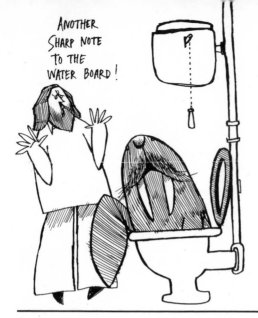

ANOTHER SHARP NOTE TO THE WATER BOARD!

thoroughly flushed through *every two years.* This, with a check on your boiler and your gaskets changed where necessary can prolong the active life of your boiler to a ripe 50 years. Think of its value then. You can sell it to the Science Museum.

Emergencies

The blocked lavatory

This can cause blind panic and appalling scenes. Ludicrous objects have obviously been thrust down your pan. Newspaper and toupés, hand luggage and God knows what. Keep a four-inch plunger on a three-foot rod (this costs about £1.25, and saves hours unbending wire coat-hangers and fishing unsuccessfully in unsavoury waters). Put the emphasis on the 'plunge up' rather than the 'plunge down'.

If the plunger fails, viz. you've given your all and so has it, and when you flush triumphantly it still wells upwards alarmingly. (There's no stopping it.) Don't do-it-yourself. You need men down your man-hole. Get the pros in.

Cisterns

These overflow. You probably need a new ball valve – so hitch up the existing one so that no water can flow in (this *is* a case for the wire coat-hanger) and get on to your plumber.

Sometimes if the cistern doesn't flush properly, the reason is quite obvious on peering in, and it's all become disconnected owing to unscientific pulling.

Blocked sink or basin

In the main, your friendly plunger should do the trick. Block

54

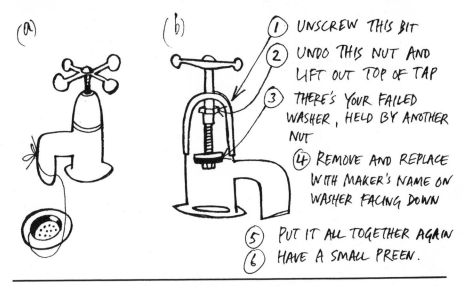

(a)

(b)

1. UNSCREW THIS BIT
2. UNDO THIS NUT AND LIFT OUT TOP OF TAP
3. THERE'S YOUR FAILED WASHER, HELD BY ANOTHER NUT
4. REMOVE AND REPLACE WITH MAKER'S NAME ON WASHER FACING DOWN
5. PUT IT ALL TOGETHER AGAIN
6. HAVE A SMALL PREEN.

the vent or over-flow on the sink before plunging. This creates sufficient pressure to budge whatever blockage you've created out of the pipe-run.

Hair and grease are the main cause of these snarl-ups. Hair-pins, too, which can't be yours, drop into the pipe-run and collect other odds and ends like cigarette-ends and the like and create a blockage in the bend.

If you've had a particularly greasy party – after washing up, run the hot tap for about five minutes. There's no need to add anything except perhaps some washing soda, hot water alone should clear it through. *But* give it a fair go because further down the pipe, say, 15 feet from your sink is very, very cold and unless you ruthlessly harass the grease with lots of hot water it will still gang up on you and create a block.

A dripping tap

Popular cause of insanity, particularly in the night.

a Temporary release from suffering as in night-time attack when amateur plumbing is out. Tie a piece of string or thread round the offending tap as in the drawing and let the string or thread lead the drips gently down into the plug-hole.

b More permanent release All you need is a new washer. (Take the old one to the shop so you get the right size.) Turn off water at the mains, and run the tap until no more water emerges and start plumbing.

One of these
days I'll be
pushing up Lilies,
or Rosie's, a Dolly's,
or Mrs Keppel's, or ...

Ed. VII

A SOIRÉE WITH A FRINGE ON TOP

Evening has fallen and you're not alone. The other needn't necessarily be the current object of your baser notions but someone from the office or a mother, who can't believe you can survive without female ministrations, particularly hers.

Anyway, it's à deux. And so are all the recipes.

THE COCKTAIL HOUR

In your interest, Gentle Reader, and with that spirit of dedicated self-sacrifice that stamps my every act, I went specially to a pub – to research the wisdom that follows. The pub was The Greyhound, to be precise, in Kensington Square, all Thackeray and nuns. The Square, that is. The pub is run by a splendidly liberated landlord, Paul Frances, who if he falls short of a working knowledge of onions, certainly knows his wall-bangers. Now in your mind you probably associate cocktails with (a) New York barmen in red bum-freezers shaking Martinis in cool, dark premises to the strains of *Melancholy Baby* on the solo pianoforte and (b) the past. They are vastly underrated, and while I may have said 'No' to a Pink Lady, I'm anybody's for a Whisky Sour or a John Collins. They provide the necessary uplift at exactly the right moment. Round about 6.30 p.m. the Cocktail Hour. (Sherry is often much misunderstood as a reviver. Gin and sherry in equal proportions could bring the colour back to Lazarus' cheek in a nonce.)

But here are some ideas mooted during a most convivial hour by that wise and good man, Paul Frances, to whom I would raise my hat, but it looks silly on paper.

Long and cool

a Campari, grapefruit juice and lots of ice.

b I can heavily recommend this one – particularly when you have no idea what you want, and the palate is jaded and needs pampering as in Asia: Vodka, American Dry ginger ale, and stuff the glass with ice and slices of cucumber.

Slightly hairier and none the worse for that

a The Celebrated Harvey Wallbanger

Mix vodka and orange juice and a lot of ice well. Stick in a glass and float a shot or two, to taste, of Galliano on the top.

b The Bahama Mama

Blend in your blender or get an old-fashioned shaker and shake with vigour (the exercise is tremendously good for you), two shots of Bacardi, one shot of Grenadine, one shot of Galliano, orange juice and ice. Pour into a glass full of fresh fruit.

c Sangrilla

Marinate some black bananas, apples, raisins, pears in a bottle of red wine or so for three hours. Add in equal proportions – Cointreau, Drambuie, and Galliano. Keep tasting. It's fun. Add a dash of lemonade for fizz.

d A Bahaman Screwdriver

Galliano and Bacardi and cold fresh orange juice.

More than a treat, a food

a The Bloody Mary

Blend or shake together vodka, tomato juice, Tabasco, Worcestershire sauce, salt and pepper.

b Red-Eye, or the Original Bloody Mary (historical note)

58

As previously advertised the old hangover cure of cold lager and tomato juice with Worcestershire sauce.

c Bullshot

Put a tin of Campbell's clear beef consommé in the deep freeze for a while – it should be liquid still when you remove it, but really cold. Add vodka, salt and black pepper, a dash of Worcestershire sauce, a squeeze of lemon.

A few more I've just thought of

Pernod. Ouzo. A *Sidecar,* which is equal explosions of lemon juice, brandy, and Cointreau. You can go on forever. Go on. Go on forever.

A Grasshopper

Quaint, but different, it is simply Crème de Menthe, fresh cream and a little nutmeg sprinkled on the top.

Zambucca. Float three or four coffee beans in it. Light it with a match and serve. Do not burn lips on hot glass.

Post-prandial drinking

This is just an excuse to give another suggestion of Paul Frances', a perfect end to a perfect evening, and it could be the start of something small.

You need two large brandy balloons. (You don't really need two large brandy balloons, but if you have them – excellent – if not, two pint pots or what you will. If you must go around nicking pint pots from pubs, at least have the grace to fill them before leaving. It looks less suspicious.)

Put in two dollops of firm vanilla ice cream, sugar, brandy, a spoon so that you can fill up with hot black coffee. Wo Ho Ho!

THE ENTERTAINING OF LADIES

IN THE KITCHEN (FIRST ANYWAY)

It's a strange irony that it should be men of all people who turn out to be the finest chefs, hairdressers and couturiers. Particularly strange in that 98 per cent of the male sex can't cook, manage a straight parting or design a white handkerchief. Albeitmoresoever, the one art we can dabble in is cooking, and to lure a young lady back to the Happy Hunting-Ground, cooking is the one to concentrate on. It's a fatal lure, sir, and beats etchings by a good length and a half.

I never did like cooking. I'm better now than I was, and, my word, when I've read this book, I'll know a thing or two. (I

started reading it the other day but when I flipped quickly to the end to see whodunnit I realised I'd only got this far and the exercise was pointless.)

Firstly, I couldn't understand the usual cook-book recipes. You're steaming through some exotica and they hit you with a phrase like 'mull over for a while' or 'relish' or 'carp' or 'caper' or 'fritter away on a slow burner', and, secondly, I am bone lazy.

There are certain things I can handle, however, for instance *Easy Spaghetti or Noodles.* Within this recipe, and prior to getting down to the real nitty-gritty of low cuisine, lies a cautionary tale.

Purchase a box of Italian noodles – tagliatelli or the like.

Purchase a clove of garlic. (There's always one around anyway. There it is, down the waste disposal unit and my heavy hand won't fit. When will they make the plug-hole large enouth?) *Useful Hint:* If you have a waste disposal unit it is invariably stiff with teaspoons. This is what wire coat-hangers are for. Unbend a pair and use as pincers. I still think that under the Sex Discrimination Act, the plug-hole should be larger.

Purchase a clove of garlic.

Butter.

Right. Boil some water in a saucepan. Toss some salt in. (What so you mean – *some* salt! – How much salt?) About the size of a heaped 5p piece. (Thank you.)

When the water is boiling, drop in as much pasta as you feel like eating. If it's spaghetti feed it in and ease it round the pan as it softens. (You may think that's elementary. I used to break it into thirds to get it into the pan.)

After about 5-10 minutes it should be edible. Keep pulling bits out and checking. It becomes less disgusting as you near perfection.

Grab a sieve, and juggle pasta between a good-sized bowl and the sink, until all the water has been got rid of. Put pasta in bowl, not sink. Throw in lump of butter about the size of a packet of 20.

Seize your garlic crusher, peel and crush two cloves into the bowl. Toss the lot about with a wooden spoon.

This is where in my case disaster struck. I was so near. Parmesan cheese, I thought. A good thought, and sure enough there on the top shelf a tiny container labelled 'Parmesan cheese', which I had inherited from my mother, when she left for sunnier climes, bequeathing me a life-time's collection of kitchenabilia. With gay abandon, I unleashed the Parmesan. Perhaps my reflexes had been momentarily dulled by the drink I always find so necessary when dabbling round the stove. Suffice to say, in the very spot where Parmesan cheese would have looked so well, viz. all over the spaghetti, lay a lifetime's collection of Hundreds and Thousands. All melting rapidly.

Never one to surrender completely. Not at once, anyway. I put the spaghetti under the tap whence it emerged still brightly coloured. Yellow, green, blue and red. I reheated it, put in more butter, added more garlic and served it. No one complained about the taste. Many remarked on the psychedelic nature of the spaghetti. Some wondered. A couple marvelled. So –

Psychedelic Spaghetti
Spaghetti, butter, garlic and Hundreds and Thousands to taste. (The Hundreds and Thousands are not compulsory.)

Thereby I also discovered that there are certain basic bits of equipment that are essential:

Things to have in cooking area or kitchen
Wooden spoon(s)
Garlic crusher
A blender or liquidiser, or egg whisk at least
A casserole
A roasting tin
A pie dish (this can be half your casserole)
2 saucepans (steer clear of cheap aluminium)
Non-stick milk pan
Tin foil
Paper towels
Good knives, for bread and carving and a good little one for tomatoes and onions
Wooden board for cutting bread, meat and rolling pastry, etc.
Sieve
A pair of kitchen tongs
Oven glove
One of those things the name of which I've never known for getting fried eggs out of pan
Washing-up machine
Pepper mill
Salad shaker
Decent stainless steel grater
Kitchen scissors (strangely essential)
Paper plates
Paper napkins – you can use paper towel
Tin opener
Corkscrew
Bottle opener
Plastic measuring jug
Plastic bowls of varying sizes
Large flip-top rubbish bin with plastic bag inside

Chefs that pass in the night

She arrives for dinner – you invited her you fool – and you're still up to your armpits in sheep's carcasses and cabbage leaves. No bad thing, it shows you care. So, what to do with her meanwhile:

1 Put her in the most comfortable chair, with suitable alcoholic drink.

2 Have the telly on, with the sound off, so that if she seems at all interested in what's going on, you only have to turn the sound up, whereas saying 'Would you care to watch telly?' sounds immediately as though you have nothing better to offer. The ball is more in her court.

3 Have the current *Playboy* or *Mayfair* or *Penthouse* or one of those lurking, ladies love to see inner workings laid bare. It's all a spicy start to the evening, and during the first drink, you could be discussing heatedly 'Masturbation – Yes or No?'.

4 Never be stuck for casual banter. There's always a conversation piece in the papers. 'Did you see about the bloke who said he was researching into calorie-counting, and was sent down for six months for weighing girls' breasts on the kitchen scales?' Not only have you a fascinating gobbet, but can easily turn the conversation towards breasts or depths of depravity.

The ice will be further broken if you are in the kitchen and she's elsewhere, and both are obliged to shout. Even if you heard her confess to mild interest in flagellation, it does no harm to shout, 'Sorry. Missed that!'. It brings her further out into the open.

5 Leave any shirts, etc. you need mending under the cushion of her chair. If all else fails you can remove them with an apology, point to the dangers of the needle you've been wrestling with, and leave her to do it.

6 Pour her another suitable drink.

Now what?

Well, there you are, noisy reader, in your kitchen, bellowing the idle chat, with the main course under way. Now is as good a moment as any to take stock. Why are we here? Where are we going? What's it all about, Alfie? Peer into your *storecupboard* and see if you have roughly the following. There, for instance, is the starter you've forgotten all about.

See still-life of storecupboard over.

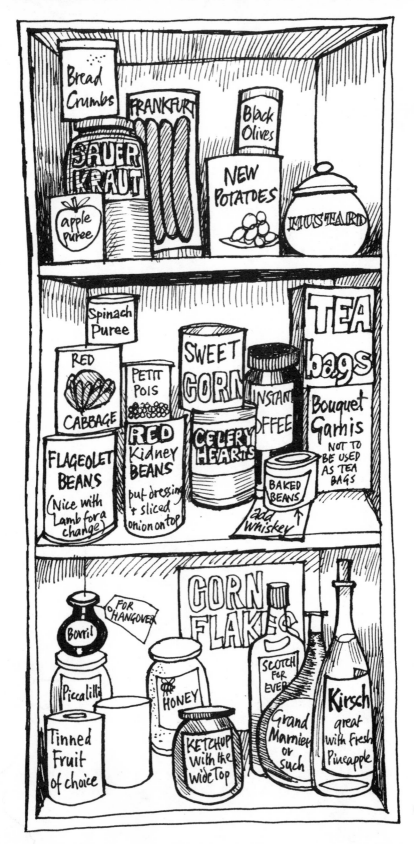

65

Instant notions which leap from this storecupboard:
SOME STARTERS IN THE RACE OF LIFE
All quantities are for **2** people.

Tuna Fish

Just drain off the oil. Stick tuna on some lettuce. Pile thinly sliced raw onion on top. Give it a squirt or two of lemon juice.

Sardines

Add black olives, artichoke hearts and some slices of tomato and a hard-boiled egg. Bang on a dollop of mayonnaise.

Stuffed Vine Leaves

Tip out contents of tin onto plate.

Humous

Empty tin of humous into a bowl. Stir in crushed garlic, salt and pepper. Then rather more deliberately, stir in plenty of olive oil bit by bit, adding a squirt of lemon juice each time.

Taste it constantly – and keep thinning it until you've hit the right consistency. Thick enough to dip fresh bread into (a lot of 'delis' now sell packets of Greek-type) and not so thin that it drips all over your tweeds.

Taramasalata

Seize your bottle or jar of smoked cod's roes which should be ready-pasted. Pound up with four slices-worth of fresh white breadcrumbs. Mix until you have a nice, smooth cream with plenty of olive oil and lemon juice. Your blender is good for this if you have one. No need for salt as you will discover. Put in a shallow dish and sprinkle on chopped mint or parsley. Dip into it with French bread or even better Greek bread if you can get it.

Starters from storecupboard plus other items:

Les Crudités

Which is not a bad name for a male striptease artist. This makes a fine and simple starter, and gives the impression you've travelled. Good raw vegetables and a garlicky mayonnaise. Cleaned-up carrots, the dissected heart of a cauliflower, celery, radishes, a spring onion or two, as long as you both have one.

How to make a garlicky mayonnaise
Use your blender. Put in 2 egg yolks, 2 tablespoons of wine vinegar, 1 tablespoon of French mustard, salt and pepper and 1 clove of garlic, crushed. Start blending in a low gear and gradually add $\frac{1}{2}$ pint of olive oil through the cunning hole in the top until it strikes you as being very like mayonnaise.

How to make non-garlicky mayonnaise
Leave out the garlic.
A cure for garlic If exuding fumes not dissimilar to a Saturday night at the Gâre du Nord – chomp on parsley.

Hors d'oeuvres

Select any happy combination of the following. Not the lot at once, of course.

Salami and sliced cucumber Negroni, if asked for by name, is delicious and about the circumference of the average cucumber, so mount one on the other and it makes a delicious mouthful. Eat the skin too, be a devil, removing it is a waste of time.

Mortadeller cheese, pâté or ham. Pickled gherkins.

Hard-boiled eggs – Slice in half. Remove yolks and, having worked them into a paste with mayonnaise, replace and pepper (black or cayenne).

Tomatoes, thinly sliced, salted and vinaigretted.

A tin of *artichoke hearts or asparagus tips* with a vinaigrette dressing stand up on its own as a starter.

Tart up your plateful with lettuce.

Note: If you open a tin of asparagus at the bottom you won't totally ravage the delicate bits up the sharp end.

If feeling flush and you're near a good deli, you can always commence with *Parma ham* and *melon, or smoked salmon,* lemon and good brown bread and butter.

ALTERNATIVELY SOME AMAZING SOUPS
(see your storecupboard)

Celery Heart Soup

Grasp tin of celery hearts. Bung the contents into the blender. Dilute a stock cube in hot water. (How much? you will ask. As much as you like, snivelling dolt, depending on how thick you like your soup.) Blend it. Heat it gently in a saucepan and Woo-hoo.

Sweetcorn Soup

Tip tin of sweetcorn into saucepan. Add $\frac{1}{2}$ pint of warm milk. Add salt and pepper. Cook for 10-15 minutes. Blend it or sieve it. Add a lumb of butter and some cream, and there you are.

Spinach Soup

Dilute a chicken stock cube in $\frac{3}{4}$ pint of water. Pour into a saucepan with a tin of spinach purée. Add salt, pepper, nutmeg and lemon juice. Heat this heady concoction thoroughly. Beat

WAITER! THERE IS A HERR IN MY SOUP.

---

up an egg in a cup. Stir a bit of the soup in with it. Then hurl all back into saucepan and stir for another minute. *Don't let it boil* otherwise the egg will cause curdling.

Green Pea Soup

Blend a large tin of green peas. Add some mint, a little caster sugar, salt, pepper, and enough milk to make it creamy. Heat it up, and do not hesitate to throw in some chopped ham, croutons too if the mood's upon you, prior to serving up.

All right – how to make croutons Cut some thick slices of bread. De-crust them. Cut into squares. Heat oil, butter and garlic in frying pan and fry bread till golden, as opposed to black. They don't half perk up a bowl of soup.

Cold Cucumber Soup

Mix $\frac{1}{4}$ pint of milk, $\frac{1}{4}$ pint of single cream and $\frac{3}{4}$ pint of yoghurt. Peel the cucumber and cut into matchstick-like pieces. Add salt, pepper, chopped mint and well-crushed garlic. Stick in the fridge for a couple of hours and just prior to serving, gaily besprinkle with mint.

A Salty Note: There comes a moment in the manufacture of a soup or perhaps a stew when in the excitement you overdo the salt in no uncertain manner. What can you do to prevent it from tasting like the Dead Sea Scrolls? Throw in a load of uncooked potato and boil it up for five minutes. Remove the potato and breathe again.

68

THE MAIN COURSE
(a pretty big moment for you too)

A Roast is a Roast is a Roast

There are those of you who will scoff. Escoffier would scoff. Scoff off. This is addressed to all those who like me have never before roasted a dead thing.

Lamb is favourite for beginners:

1 It matters less if it's over-cooked (i.e. cook has overdone the Dutch Courage or Dutch Worthington E).

2 It matters less if it's under-cooked (blind panic or lack of faith).

So herewith for learners:
One Roast Lamb Dinner for two

Before you begin:

Approach your friendly, neighbourhood butcher. A good man to keep in with. He'll tell you how much you need, and advise on how long it'll take to cook.

You: 'You are a wise and good man, butcher. Your finest shoulder of lamb, please. Three pounds or so and a lean one too.' (You'll know if it's lean by looking at the blunt end. Raise the eyebrows and shake head if it's white and fatty.)

Lay hands on:

A number of potatoes
Some frozen veg. Broad beans or green French beans are good.
Peas or carrots or both
Butter
A packet of Bisto or the like
Garlic – one clove will do
Jar of mint sauce or redcurrant jelly or both (in your storecupboard?)
Salt and pepper
A bottle of sunflower oil or corn oil or other cooking oil.

Arm yourself with:

1 roasting pan
1 saucepan for spuds and frozen veg.
1 wooden spoon
1 tablespoon
1 carving knife
1 fork
1 oven glove
1 watch or alarm clock.

WHO ARE YOU TO TALK?

Good news for chauvinists
English lamb is leaner and more expensive!
New Zealand lamb is fatter and cheaper.

① Clean Joint with a paper towel and put it in Roasting Pan

② Melt a pat of Butter (about the size of a packet of 10)

③ Slap Melted Butter over Joint. Spread evenly with back of spoon or old toothbrush

④ Slice up clove of Garlic and poke slivers into crevice round bone

⑤ Throw rest of Garlic in round and under Joint. You can suck them later like boiled sweets.

⑥ Put enough Cooking Oil around Joint to coat your Roast Spuds thinly

⑦ 1¾ Hours to Go Switch on Oven Electric – 450°F Gas – No 8

⑧ DON'T PUT JOINT IN YET!

TWIT

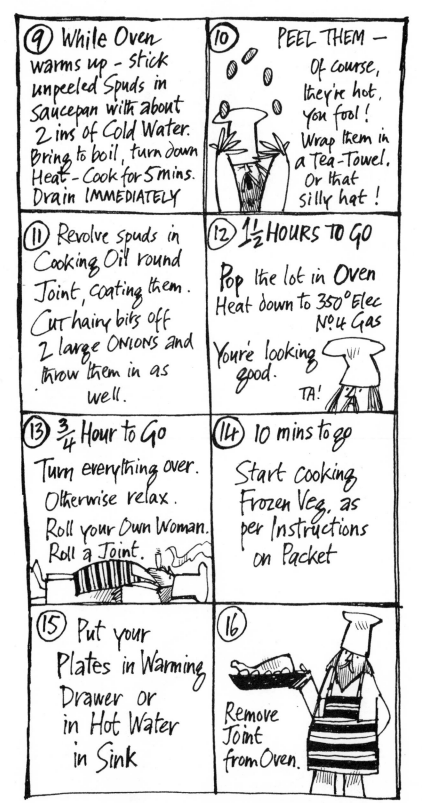

⑨ While Oven warms up - stick unpeeled Spuds in Saucepan with about 2 ins of Cold Water. Bring to boil, turn down Heat - Cook for 5 mins. Drain IMMEDIATELY

⑩ PEEL THEM — Of course, they're hot, you fool! Wrap them in a Tea-Towel. Or that silly hat!

⑪ Revolve spuds in Cooking Oil round Joint, coating them. Cut hairy bits off 2 large ONIONS and throw them in as well.

⑫ 1½ HOURS TO GO

Pop the lot in Oven Heat down to 350° Elec N°4 Gas

You're looking good. TA!

⑬ ¾ Hour to Go

Turn everything over. Otherwise relax. Roll your Own Woman. Roll a Joint.

⑭ 10 mins to go

Start Cooking Frozen Veg. as per Instructions on Packet

⑮ Put your Plates in Warming Drawer or in Hot Water in Sink

⑯ Remove Joint from Oven.

(17) YOU HAVE FORGOTTEN OVEN GLOVE!

GO BACK 16 SQUARES

(18) DECANT Joint, Veg.etc onto Warm Plate and put in Warming Drawer — or Turned-off Oven

(19) DON'T POUR FAT DOWN SINK. PUT IT IN A BOWL — LEAVE A BIT IN PAN FOR (20)

(20) Make GRAVY as per Instructions on Packet — BUT heave in a glass of RED BIDDY

VINO

(21) Serve up in Kitchen. Saves time and Washing-Up. Remove Onion skins

(22) Throw all your Accoutrements into Sink — Leave to soak in Hot Water and Washing-up Liquid.

(23) You have Finished

I beg Your Pardon

(24) You have finished

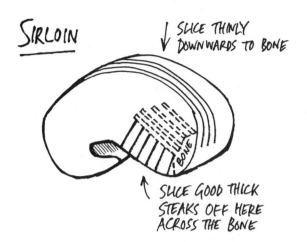

SIRLOIN

SLICE THINLY DOWNWARDS TO BONE

BONE

SLICE GOOD THICK STEAKS OFF HERE ACROSS THE BONE

Roast Beef

This is exactly the same exercise as roast lamb, but roast it for 20 minutes a lb + 20. It shouldn't be over-cooked, although of course tastes vary and who am I to quibble if you prefer charred remnants. Perhaps the beast wanted to be cremated.

It will need slightly more attention than the lamb while in the oven – quite frequent slappings of fat all over it.

(Incidentally for this you can buy a remarkable implement not unlike a douche for a very tall lady. You suck up the fat and squirt it. It's quite fun, and not habit-forming.)

There's nothing like *horseradish sauce* with beef. Keep a bottle of it in your storecupboard and stir into it a blob or two of cream before you serve it.

Roast Pork

Now this should *never* be under-cooked. Vile diseases lurk. Lock-jaw, foot-and-mouth, Dutch Elm even. So aim for 30 minutes a lb + 30.

I do enjoy a nice piece of crackling.

I'M NOT AVERSE TO IT MYSELF

73

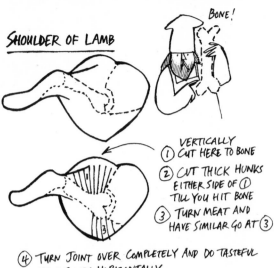

SHOULDER OF LAMB

BONE!

VERTICALLY
① CUT HERE TO BONE
② CUT THICK HUNKS
EITHER SIDE OF ①
TILL YOU HIT BONE
③ TURN MEAT AND
HAVE SIMILAR GO AT ③

④ TURN JOINT OVER COMPLETELY AND DO TASTEFUL
THIN SLICES HORIZONTALLY

Ask your new friend the butcher to score the fat for you. Butter as for the other joints you've mastered, and liberally sprinkle with salt. If when you put the pork in your roasting tin, you slip some sort of rack under the meat (there's doubtless something suitable floating round the oven), this will keep your crackling crisp.

Serve with *apple sauce* (from the tin in your storecupboard of course).

CARVING

This is a bit of a myth. After all it's up to the individual to chop it into mouthfuls anyway, all you're doing is having first hack.

However, for reasons I don't understand, it seems to be a male preserve, and you do find yourself on occasion faced with a joint, a knife and a fork. A sharp knife is worth any expertise. If you turn the beast over a few times, and prod it knowingly with the knife, you *(a)* look as if you know what you're about and *(b)* give yourself some idea of the geography – soft terrain, bed-rock and the like. Try and make the slices as large as possible. This impresses, if you seek to impress.

Here is a rough guide to a *shoulder* of lamb and how to dismember it.

With a *leg* it's almost exactly the same principle, except that you won't have that large saucer-like shoulder-blade to deal with.

Beef

This is invariably simplicity itself, unless you get ambitions and buy a sirloin.

74

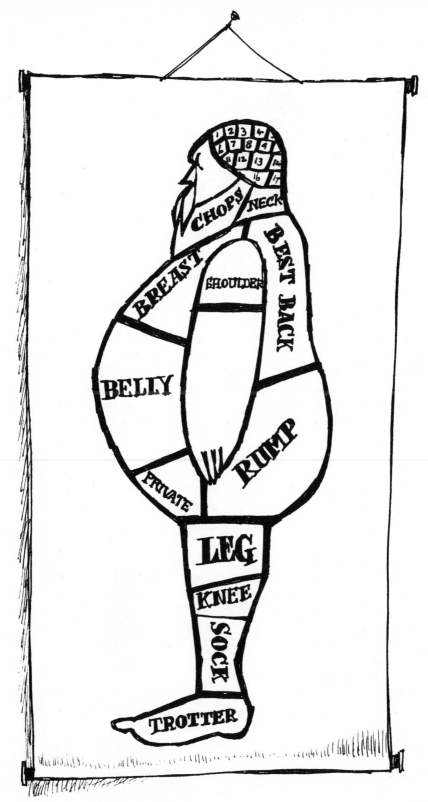

75

Chicken

A chicken should almost fall apart anyway. Wings off first, then the legs (if you want to spread a bit, separate the thigh from the drum-stick), then simply slice the breast up. It reads like Jack the Ripper at work on your guardian angel. Don't forget, and some do, the two delicious 'oysters' under the bird, to my mind the *bonne bouche*. Always enquire if anyone wants the parson's nose. There are those.

Turkey

Identical to chicken, but as they are larger it's best to take slices off the legs and wings rather than serve up the monster limbs in toto.

What to do with a sudden chicken

After all you might suddenly become the proud possessor of a chicken. In anticipation of this event therefore, buy yourself one of the great cheats. *A chicken brick.* This is a chicken-shaped unglazed earthenware pot.

All you do is put the chicken in it, brush the bird with oil, salt and pepper and then either

a sprinkle rosemary all over the top

b put sliced mushrooms and a squirt or three of lemon inside the chicken. Squirt outside too – it makes a grand sauce; or

c squeeze as much garlic as you can stand all over.

Put the brick into a cold oven (or conversely sell it to the Tate Gallery). Turn the heat to 500°F (260°C or Gas No. 9) and cook for $1\frac{1}{2}$ hours. It cooks in its own juice. If your oven is blessed

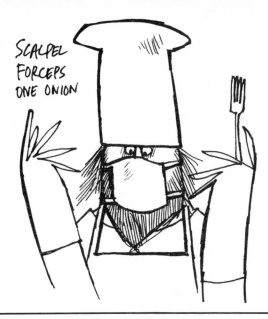

SCALPEL
FORCEPS
ONE ONION

with an automatic timing device, you can of course prepare it all in the morning and come home to it. The only problems here are that you may forget all about it, pubs being what they are, and some may find it beyond them to stuff mushrooms up the back of a chicken in the cold light of dawn.

Eat it with rice or noodles, or fresh bread and butter.

Further adventures with a chicken brick

You can roast a loin of pork in it. Make sure it fits your brick. Three pounds should, and that takes $1\frac{1}{2}$ hours to cook if you put it into a cold oven again set at 500°F (260°C or Gas No. 9).

Life with a casserole

This makes life very simple. Admittedly it can't handle your VAT, but as a method of cooking it is remarkably trouble-free.

Buy a heat-proof glass or earthenware casserole. These are cheap. The French ones are best, but expensive. This is very true of their women, of course. The French refer to them as *daubes* – the casseroles, that is, not their women, God bless them. Here is but one thing you can do with it:

Casserole of Beef

Buy 2 lb of *topside of beef* in one piece. Put the beef in the casserole and cover it with slices of fat bacon. Add salt. Add herbs. Parsley, the ·odd bayleaf, a bouquet garni. (What is a bouquet garni? – a herbaceous tea-bag.)

Smack in a glass of whatever wine you have in mind. have one yourself. But do remember that the Galloping Gourmet ended up with religion.

Make a good glue with flour and water and thus stick the lid to the casserole. This prevents steam and flavour from escaping. Put in a slow oven. 275°-300°F (140°-150°C or Gas No. 1-2) for 4-5 hours. After 4 or 5 hours take it out, and serve it with the sauce it's made for itself, noodles and 3 or 4 bottles of rude Red.

There are countless variations on this theme. Go mad.

Now that you are feeling extremely cocky a whole new world has now been opened up to you. Amaze yourself further.

Make a steak and kidney pie

Equip yourself with the following:

> For two people – buy $1\frac{1}{4}$ *lb of*
> *steak and kidney.* Ask your
> friendly butcher to chop it up
> for you – particularly the
> kidneys – I'm not that
> squeamish but chopping up
> kidneys does smack a little
> of amateur surgery.
> *A packet of ready-made,*
> *   frozen puff pastry.*
> *Flour*
> *1 onion*
> *1 bayleaf*
> *1 clove of garlic*
> $\frac{1}{2}$ *stock cube*
> *and a clean milk bottle?*

On the day before, or on the morning after the night before, cook your meat. It needs time to cool. So will you.

How to cook the meat

Roll your meat about in some flour to which you've added salt and pepper. You can do this in a bowl or your hat if you like. Slice an onion thinly. Put the meat, the onion, a bayleaf and the garlic into your casserole. Having mixed the half cube of stock with sufficient hot water to almost cover the meat, but not quite, cook it all, with the lid on, at about 300°F (150°C or Gas No. 2) for 3 hours.

Jacket potatoes which are favourite with it: With 2 hours to go, scrub some large potatoes, stick a fork in good and deep all over to let the heat in, and put them on some tin foil near the top of oven.

Now to make the pie:

78

① 1¾ Hours to GO

Take FROZEN PASTRY
out of Freezer –
(It takes an hour
to defrost)
"A WATCHED FROZEN THING
NEVER DE-FROSTS"

② 1 Hour to GO

Put the Oven on.
425°F or Gas № 7

③ Sprinkle flour on a
flat surface to stop
sticking as you roll
out your Pastry to
thickness of 10p-piece.

④ Put meat in Pie Dish.
If it's not full up – put
an EGG-CUP in the middle
upside-down to stop your
Pastry sinking

⑤ The Pastry should
be roughly the shape
of Pie Dish – but
larger

FLY'S-EYE VIEW

⑥ Damp lip of Pie Dish
with milk. Use finger.
Stick on strips of Pastry

Thus-
ish

⑦ Damp the strips and draping Pastry over Bottle - lay it over Dish.

⑧ Press down edge with a Fork and cut away the untidy bits with a sharp knife

⑨ Make a slit in the top to let steam out

⑩ With surplus Pastry - tart up Pie

SOD OFF

with tasteful message

⑪ Brush pie all over with Milk — (or Beaten Egg for that glossy look)

⑫ 45 minutes to GO

The precise time required to Cook it at 425°F or Gas 7

You've only bloody done it again!

I USED TO BE QUITE BIG DOWN UNDER

Things to do with steak
Steak Sandwich
Fry a thin piece of steak in butter or olive oil, put it on a piece of bread and butter or toast and coat in tomato ketchup. Add as much Worcestershire sauce as you like to the ketchup. It performs tiny wonders for it.

Steak in the Australian Manner
Fried, with two fried eggs on top, electric ketchup and a tin or three of ice-cold Australian lager.

Pepper Steak
Having lacerated the steaks lightly with a sharp knife, rub it all over with garlic. It will feel better for this.

Crush some black peppercorns (about a heaped teaspoon per steak – more if you're a pepper perve) on the breadboard with a meat-basher if you have one. Or put between greaseproof paper and placing said on floor, leap up and down on them. Look at the state of the floor. Buy a pestle and mortar.

Rub the poor pounded peppercorns into the meat vigorously. You will feel better for this.

Fry your steak in butter and oil. If you haven't burnt the butter pour it over the steak, and serve with jacket potatoes. Salad, too, if you like. (See a Salad page 88.)

How to deal with veal
4 escalopes of veal $\frac{1}{4}$ pint cream

Salt and pepper the escalopes. Melt some butter in a pan, and

81

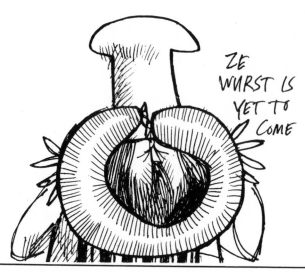

ZE WURST IS YET TO COME

keep on revolving the escalopes merrily till brown. Give them a prod or two and when tender, put them in a warm diship.

Heat the cream in a separate pan. When hot, pour the cream into the recently vacated veal pan and stir round with a wooden spoon and then pour the vealy, buttery, creamy sauce over your escalopes.

Add green peas, mushrooms and new potatoes and announce loudly *Escalopes de Veau à la Crème* which makes it sound more difficult than it was.

Poached Sausage with Hot Potato Salad

(You can always boil up some Frankfurters, empty some salad cream over a tin of new potatoes and run it up the flag-pole but this way's better. Believe me.)

The poached sausage

Buy one of those large Continental sausages that loops back on itself in a circle like a suicidal worm. Or simply ask the man at the deli which Continental sausage he, personally, being clearly a man of taste, would poach – the mood upon him.

Poach is the operative word – as in Lincolnshire. Not that a Continental sausage was his delight on a Friday night. Do not, on any account, *boil*. The sausage should be well-submerged in gently bubbling water. For a one-pounder, allow an hour. For any other weight, work it out yourself.

The hot potato salad

Boil the spuds in their skins until they're just cooked, but short of that moment when they disintegrate.

82

Peel them.

Cut them into nice, thick rounds.

Cover them in vinaigrette dressing.

('How do I make vinaigrette dressing, chief?' you enquire. That should have read – 'How do I make vinaigrette dressing, chef?' but printers are only human. How? Am I not your friend? See page 89).

Meanwhile back at the saucisson

Now put the sausage on a plate and decorate with hot potato. Don't let the potato get cold. Apart from anything else it makes a mockery of the happy phrase 'hot potato salad'. Add lettuce, if you like, you have the dressing in the fridge. Add a nice pot of mustard. Dijon. Sometimes fancy, it's a pleasant compromise between them and us. Who loves you, baby?

FISH

Very well, I confess. On the occasions when the brain cries out for fish (and I lean more towards the Jeevesian theory that fish repairs brain damage and indeed enhances its performance, than towards those who say this is a load of cobblers), I either buy it frozen, already be-sauced, or go to a good fish and chip emporium, greatly improved in London since traditional English methods have given way to the Greek.

Still you may wish to take on a fish dish either as a starter or a main course and as at the time of writing the Cod War still rages (whose fish finger on the button? and the like) here is a recipe that involves red or grey mullet or the good old mackerel.

Pesci Arrostiti e Freddi

Guess who invented this? Its main joy is that it's cold so you can run it up any time and keep it in the fridge.

4 fish (red or grey mullet, or mackerel)
¼ pint olive oil
2 bay leaves – shredded
1 desertspoon each of chopped parsley, chives, chervil, a little thyme and fennel, cayenne pepper, salt
1 lemon

Ask the fishmonger to clean up the fish when you buy them. Mongering fish is *his* business.

First, switch on the oven to 375°F (190°C or Gas No. 5). Sprinkle the fish with cayenne pepper. Put them in a shallow oven-proof dish.

83

FREDDI

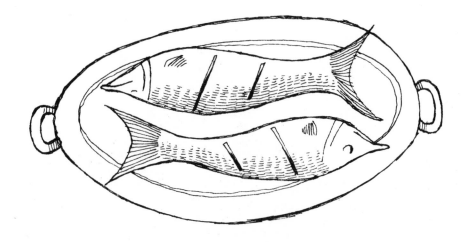

Make some long cuts in one side of your fish and pop the herbs in the slits, with some salt.

Pour half the oil over them. Squeeze half a lemon over them.

Stick them in the pre-heated oven, nice and near the top, so they can tan. About 25 to 30 minutes will suffice, depending on their size.

Leave them to cool, and just before they're quite cold pour the rest of the oil and lemon juice over them. Put in fridge. They should be served very cold with potato salad.

VEGETABLES

Now in the main, because I'm as indolent as ye, I can see no point in pursuing vegetables further than the frozen food section of the local supermarket, or tins. Unless you grow your own which is another story. (See – another story concerning growing your own vegetables pages 137 to 139.)

However, there is one vegetable with which you might care to dabble. The long-suffering product of the soil the English abuse so freely, usually by boiling it to perdition. It can be no coincidence that while over here it is used as an insult, in France it is a term of endearment –

Gentlemen, the Cabbage

The cabbage plain and simple

Shred it fine, croon to it, wash and drain. Cook in a very little boiling salted water with a slice of onion and, if you like, some caraway seeds. *Do not overcook!* This is why the Empire crumbled. Let it be crisp and cabbagey.

YOU NEVER KNOW
WHEN A
VEGETARIAN WILL
DROP IN

86

Sweet and sour cabbage

This is particularly useful if you have a pair of long-forgotten apples festering in the kitchen, although you may consider it more to your liking simply to buy them.

2 sour apples
1 red or white cabbage
vinegar or lemon juice
2 tbsp flour
4 tbsp brown sugar
2 tbsp oil

Finely shred your cabbage then peel and slice your apples (or vice versa). Sprinkle with salt and pepper. Heat the oil in a saucepan and add cabbage and apples. When well mixed, just cover with boiling water, and leave to cook for about 10 minutes or so until they are tender.

Then sprinkle the flour thereupon, add the sugar and vinegar and mix away.

Cook for a few more minutes while the flavours merge, and you've allowed yet another cabbage to retain its self-respect.

A Vegetarian Lady

I will confess if I wasn't so fond of roast beast-of-the-field, I would be a vegetarian. I remember once putting a lady off brains for ever, by observing lightly that she was about to chomp on a mind. Your meat-eater has to try quite hard on occasion to weigh the joys of, say, ox-tail against its natural purpose and position. And when it comes to the lights and the offal, least said sooner eaten.

A Vegetarian Pie

Quickly adapt the steak and kidney pie you had prepared by making one half for her and one half for you.

Make exactly as per instructions for steak and kidney pie but fill with a delicious mixture of lightly *cooked* vegetables (e.g. carrots, beans, onions, turnips). Save the vegetable water for making stock. Add a handful of cooked haricot or butter beans, top with a layer of tomatoes. Add brown sauce made from vegetable water or stock and gravy browning.

Cover filling with pastry as for steak and kidney pie – brush with milk and bake in a pre-heated oven at 425°F (220°C or Gas No. 7) for 30 minutes or until brown.

Note: this pie can be covered with mashed potato instead of pastry and browned in the oven – then you have a *Vegetarian Shepherd's Pie*. Damned odd employment for a vegetarian – but these are hard times.

SALADS
A classic Rushton salad in the making
When making a salad there are three important secrets:

1 The ingredients should be as fresh as possible but not include any livestock.

2 They should be preferably washed and chilled well beforehand.

3 Variety and mixture of flavours is what makes it unforgettable.

Select and prepare your greens with care and discretion

Lettuces vary according to the season, like me, but try not to buy a sad, wilted looking one at any time of the year. There are cabbage lettuces, long crisp, green cos lettuce and the curly, crisp Webbs Wonder. In addition there is watercress, curly endive, and the leaves of young spinach. Lettuce too is fragile and bruises easily so handle it with loving care. It's always best to use lettuce after at least several hours' chilling in the fridge. Wash your lettuce as soon as you buy it in plenty of cold, running water, to remove all grit and bodies.

Remove any wilted leaves. Now remove moisture thoroughly by shaking in a clean tea-towel or shaking in a lettuce basket. If you use a lot of lettuce it's worth buying the kind that 'spindries' the lettuce, whirling round and round in the sink, to which it adheres with a kind of rubber sucker. It's worth it for the thrill. Now place in the 'chill drawer' of your refrigerator.

Poor sod, it's probably chilly enough already – or put in a clean polythene bag.

Some tasteful companions

Chopped, fresh *chives* add flavour to a plain green salad, so do dandelion leaves in the spring if the donkey hasn't eaten them all, and nasturtium leaves in the autumn. Thinly sliced *radishes* add colour and piquancy (I like a bit of piquancy), and chopped *spring onions, raw mushrooms,* sliced *tomato.* When tomatoes are very expensive you can still add colour by grating some *raw carrot* finely and sprinkling it first with lemon juice and oil. Other additions are grated *celery,* raw *white cabbage, endive,* or *cauliflower* that has previously been plunged into boiling salted water for about five minutes.

Getting it dressed

Give it the space it deserves in a large salad bowl or the like. Salad looks a bit sad with cramp. Shake up your salad dressing
88

well (if you have made this in quantity in a bottle – see following recipe). *Don't* dress the lettuce hours before eating or it will go limp, creased and greasy. If all your ingredients are washed clean and well dried, or prepared according to kind (as above), you should not need to toss them (i.e. turn over and over so that the dressing coats every leaf) until just before serving.

Vinaigrette dressing
(or salad dressing or sauce vinaigrette if you prefer it in the French. *La chance* would be a fine thing.)

You'll need salt, pepper, 1 clove of garlic (optional), wine vinegar and a good vegetable oil (if you can't afford olive oil).

First crush your garlic. Add salt and pepper. Add and mix oil and vinegar – oil 3 to vinegar 1. Beat all with a fork, or, put in a clean jar and shake like your sister Kate. Keep the result ever-handy in the fridge.

PUDDINGS

Afters, if you like. I once had a tragic affair with a tin of syrup pudding. If I have a weak spot in my armoury, it's where syrup pudding is concerned. If the Devil is interested in some sort of barter, then my soul is his for good syrup puddings. As long as they're not in a tin. I had one once, and the story is an explosive and sad one. Suffice to say that of the tin itself, little remained. The pudding on the other hand must have split like so many amoebae and multiplied, breeding a whole new race of syrup pudding. It was wall-to-wall, ceiling to floor. I was surrounded.

So, I suggest you try something else.

Chocolate Mousse

Of course you can buy them, I'm only seeking to impress. The advantage of this method is that it's another way of sneaking further alcohol into her bloodstream.

2 eggs (this on the 1-per-person basis)
2 oz plain chocolate (so is that)
a bottle of Tia Maria.

Put the chocolate in a basin with 1 tbsp of water. Put the basin in a pan of boiling water. Stir away with a wooden spoon, until it's all smooth and melted.

Now separate the egg yolks from the egg whites. (This requires a steady hand and no audience, and possibly some more eggs. If you crack them in half carefully, it's a matter of dextrously juggling the two halves over a basin, seeking to get the white in said basin, while retaining the total yolk in one half-shell).

Stir the yolks into the melted chocolate and add 1 tbsp or so of Tia Maria. This is best done with an unsteady hand.

Beat the egg whites until stiff. Fold gently into the chocolate mixture.

Pour the mousse into individual glass dishes, or glasses or cups or whatever and leave to set in a cool place. Only put them in the fridge if the weather's boiling. Another chance would be a fine thing.

An even more alcoholic pudding

Quick and easy, and a song guaranteed with every helping.

3 eggs for a duet (6 for a barber's shop quartet)
2 heaped tbsp caster sugar
grated rind of 1 orange
2 tbsp of Kirsch, Grand Marnier, Curaçao, Apricot Brandy,
or whatever liqueur turns you on, baby doll.

First, switch on your oven to a high heat.

Separate the eggs. Beat the yolks in with the sugar, grated orange and booze. Thoroughly thrash.

Whip the whites until stiff. Now bang the two together *con molto bravado.*

Heat your omelette pan, and pop in a small knob of butter, when hot. With delightful movements, make certain melted butter covers pan in toto. Pour in the boozy mixture. Shake the pan. Imagine you're playing the maraccas with Edmundo Ros. On no account, become immersed in this fantasy, for, at once, the outer surface will brown, and the bit nearest you to puff up like something out of Dr Who. It should be cooked in a minute.

Now for another minute put the omelette pan in the hot oven.

90

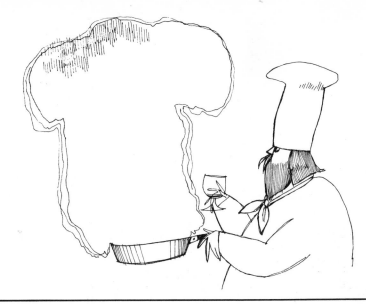

Slide the lot out onto a hot dish, folding as you so do. Folding the pudding, that is, unless it's all been too much for you.

Note: Only do enough for two at a time. If you want more, owing to more people or cries of *'Encore, chef'* from the lodestar of your life, cook more separately, but don't double up the ingredients.

Orange Sorbet
Buy 2 frozen orange sorbets
Buy 2 oranges
Place 2 frozen orange sorbets in freezer
Place 2 oranges in something tasteful like a clean ashtray.
After paeons of praise have been lavished on your roast or whatever, idly query – 'Orange sorbet? No bother.' Take the 2 oranges into kitchen and secrete. Make a fair bit of noise, sing, and return with 2 orange sorbets out of fridge, whistling nonchalantly.
Open a can of applause.

91

THE PLAIN MAN'S GUIDE TO KNOCKING OVER LADIES

You are not a plain man first and foremost. You may be a plain cook, and her eyes may not yet be sufficiently wide to have the wool pulled over them. (I'm not suggesting that you employ deceitful tricks, it's just that the only way to get the jumper off is upwards.)

Not only are you not plain. You are lovely. Say to yourself daily, 'I grow prettier by the minute'. Beauty lies in the eye of the beholder. Get a short-sighted woman, you fool and read on.

OTHER WAYS TO IMPRESS LADIES
(if your culinary efforts are not enough)

That heading was composed by a lady publisher, which makes me wary. Not that her sex has anything to do with it, of course, except that she's a woman.

A cautionary tale leaps to mind. I remember an excited lady jumping up and down and pouring out the events of the previous evening, which had finally convinced her that she'd found the lodestar of her life, the sun of her universe, and at the same time the sort of person she could take home and show to father. Apparently, she and her paramour had gone back to her place for coffee and a possible crash at earth-shifting and bell-ringing. Unfortunately, they fell into a tiff or row, without which, I've heard, no human relationship is possible, but, God, it's pleasant. The outcome was that the paramour was shown the door, and the lady went red-faced, sobbing and alone to her bed.

The paramour either consumed by lust or unable to heed the simplest of Mother Nature's warnings climbed three floors up a drainpipe and erupted into her room.

This, she told me, was by far the most impressive thing she had ever known of, and as a result she was his for the plucking. 'What a man', she thought. 'Whatever will he do next?', she said, juddering with contentment.

This is where, unbeknown to her, she'd zeroed in on the problem in one foul swoop. What *was* the poor stiff to do next? Once you've established climbing three floors-worth of drainpipe (this I think was in the Battersea area, where they are of an age and none too sound) – what next? You can quite often obtain the same effect with a well-turned bunch of flowers, old fashioned this may seem but it could still give Germaine Greer cause to pause and as a feat, is none too difficult to follow – but start behaving like the man in the chocolate commercials, ski-ing through avalanches, and jumping rapids and through fiery hoops to press a melting product into her bed and in no time whatsoever she's off in search of even further adventure and you're nursing your first coronary.

Romance is one thing but, whereas Douglas Fairbanks spent more time in sword-play or mid-air, Charles Boyer always used the front door and the minimum of sweat.

There's a great lesson to be learnt here from the activities of the weaver-bird, and to save you donning a hide and clambering through uncharted undergrowth with binoculars to peruse the little fellow at his activities, I'll tell you what he's at.

The Gentleman Weaver, as his name suggests, weaves and knots a delightful little basket of a nest and then exhibits it to the Lady Weaver of his choice. If she turns him down, he pulls his nest to pieces and starts all over again. You may think that this is a pretty non-productive bit of ornithological activity, but, apart from the basic truths contained in its enactment, thrill to the best bit and shout loudly 'Ain't Nature a hoot?' Until the moment your weaver-bird gets a 'yes' or 'no' from Lady Weaver, he doesn't tighten the knots on his nook – thus making it infinitely more simple to demolish. No pain. No sweat.

How then to impress, apart from climbing drainpipes?

Appear to be decisive. Say 'Tonight I am going to see *Ain't Nature A Hoot?* at the National, after which I shall go to a Greek restaurant in Charlotte Street, then home to Irish coffee and bed. How does that grab *you*, baby-doll?' It grabs her. It grabs her.

The old courtesies. Opening doors, flowers, walking on her outside (I don't mean a Japanese massage), holding the umbrella

94

over *her* head, helping her on with her coat, helping her off with her tights. 'Little things', as the old song had it, 'mean a lot'. In fact the old songs usually had it. *Try A Little Tenderness; The Way You Look Tonight; Tell Her You Love Her; Come Into The Garden, Maud; A Bungalow, A Piccolo And You* to name but a few.

Always take a refusal gracefully. There is no more disgusting spectacle than some man whining or ranting at some poor lady who has turned him down quite nicely for what may be excellent reasons: *(a)* he is bone ugly and smelly to boot *(b)* she has the curse of the bride of Frankenstein.

Don't bother to do anything famous. There's nothing famous left to do. They've climbed Everest straight up, the hard way, backwards, with and without oxygen or women, they've ski-ed down it; they've sailed round the world single-handed, both ways back-handed, look-no-handed; swum the Channel over and over; they've reached the North Pole time and time again, over the ice, under the ice and on pogo-sticks; won the West, gone down in the East, drifted South and danced on the moon. Just clean your teeth and give her a kiss.

(I remember Grandfather saying that the worst part of swimming from Dover to Calais was the walk to Paris afterwards in damp tails.)

Never show fear. I'm looking at a copy of the *Daily Mirror* now. 'Starts today – A provocative new series on the new breed of women. The She Male. She's bold, brassy and belligerent' and labelled later – 'The Masculine-Aggressive Female'. *Try* not to show fear, then.

Be a bastard on occasion. Oh, I've been a bit of a bastard in my time. The appalling thing is they seem to like it.

What about those liberated ladies?

Surely all this is a fraction old hat, you may say, opening doors and conquering the north face of the Aga, so – what about Women's Lib.?

I am old hat. I wear an old hat. But I still doff the old hat and until I'm gang-banged end-to-end by a squad of heavily armed, and probably heavily legged, hard-line, dead-in-the-wool, extremist, activist Libbers, I shall continue so to do.

Seriously though, and notice how I lean forward, oozing sincerity, like a politician evading an issue raised by Robin Day, the times they are a-changing.

Nowadays I can accept a drink from a lady in a pub without

blushing. They can buy the tickets for the movies, without my feeling like a gigolo of the old school.

Play it by ear, sir. If you've offered to buy her dinner, and when the bill comes, she is suddenly militant, tell her she can buy the next one. Suggest the Caprice.

If she says 'Tonight I'm going to see *Ain't Nature A Hoot* at the National, after which I shall go to a Greek restaurant in Charlotte Street, then home to Irish coffee and bed. How does that grab *you*, baby-doll baby?' Be grabbed, baby-doll baby, your luck has just changed.

If you've just met at a party, *give her your telephone number, and don't ask for hers.* She is therefore, if a Libber, obliged to ring, or her New Womanhood is threatened.

Enjoy the new equality. At last, it's our turn to lie back and think of England.

L'AMOUR LA MERRIER

It is doubtless fair to say, and poets have probably sung about it, persons hurled themselves off bridges about it, husbands become used to it, monks thanked heaven they don't suffer from it, that in every male life there is one lady, and only one, that turns your bowels to water. That scares hell out of you.

How strange that the poet, the suicide, the husband, the monk all burble on about the heart (the way to which is the stomach).

Where men and women meet isn't heart or head, or limbs or groin, it's the bleeding bowels. The guts. The ulcerian region.

It's not the pressure of business that causes silly varicose-faced men to keel over during their third luncheon – it's the little woman. Had Hitler changed sex, he would have been a little woman.

'There is no such thing as love', said the man in the corner. 'If God had believed in love, he'd have thought of better rhymes than glove, turtle-dove, above, and shove. At least in French *amour* rhymes with almost everything that doesn't rhyme with *cheri* or *couchez* or *lit*. But then God was English.'

I wasn't saying a word.

'If God had intended man to marry, he'd have married himself. But, you can pore through the Gideon Bible in your lone hotel room (those nights when you haven't a book or a newspaper and you've read the menu, the services they provide for you, the washing instruction labels on your clothing), and there is not a solitary mention of a Mrs. God', said the man in the corner.

'There is a theory that God is a woman', I suggested, 'which explains the mess and makes the Holy Ghost's behaviour questionable to say the least'. Then for no reason at all I thought I'd change the subject. To sex!

EVERYTHING ELSE YOU ALWAYS WANTED TO KNOW ABOUT SEX BUT CAN NEVER FIND IN THE INDEX WHEN YOU WANT IT

Sex, there I said it, and you never thought I would. Tell us the secrets, you cry, reveal all. It's not much fun a bloke telling you what to do. I know you, and all your fantasies come true when a *lady* tells you what to do – as she huskily mouths, 'Put on your sou'-wester, Arthur, and your whale-bone leg, it's *Moby Dick* tonight.' Curl up with Anna Raeburn, she's the expert at it. Now there is wisdom, housed in the loveliest of heads. I warn you I'm leaving myself out of this bit. I'm not waving my sexual antics and disasters about with you watching. (You're safe, Amanda, nary a word shall pass my ball-point about the night your contact lenses flew out on impact, and love died as my naked knees crushed them into the carpet.)

However carefully you phrase the history of your sex-life, you're bound to emerge as a boaster, a braggart, a liar or a laughing stock, apart from which you've written it down and it can be used in evidence against you. I am *not* Frank Harris. ✳

'Though, I must admit', he once vouchsafed as the Café de Paris juddered to a sudden silence, 'if Will Shakespeare had asked me, I would have found it hard to refuse'. This is from the same stable as a Lord Curzon asking some duchess or other

✳ *I AM!*

EXCUSE ME - I'M WRITING THIS BOOK...

... FOR THE SINGLE MAN

whether she'd bed down with him for a million pounds and saying upon her eager nod, 'Having established the principle, now let's get down to the hard bargaining'. Or something. Which brings us to a basic question – what price a poke? Would Frank Harris have succumbed having established the principle with the bard, to any other playwright or poet from Marlowe and Bacon to John Betjeman and Ray Cooney. Did Oscar Wilde's eyes light up? How much was Curzon obliged finally to fork out?

In the 1950s, when I first shot into puberty over a photograph of Ava Gardner, life-size, I recall, from *Lilliput*, one had far less ladies than you had hot dinners. But times they are a-changing and nowadays the ratio is roughly 50-50.

I remember Barbara Cartland writing then in her book *Wedding Etiquette,* of how rude it was for the gentleman to turn over immediately afterwards and go to sleep. Even more grotesque, and far from nice, if he failed to say 'thank you'. God in Heaven (name and address supplied) in those days if ever I'd got to an afterwards I would have bowed, scraped, raised my hat, showered her with daffodils and thanked her till my lips were chapped. Never have a *belle dame sans merci*, as the French have it, and they have it a damned sight more often than we do.

Those were the Doris days, warm handshakes on Grace Kelly's doorstep. Maidenhead revisited. Virginity was rife.

Sex education then was drawing maps on the backs of rabbits, and gathering what you could from the serialisation of *Forever Amber* in the *Sunday Dispatch.* Then *Lady Chatterley's Lover* burst upon us from under the counter at W.H. Smith's, and there

99

. . . ON HOW TO PULL LADIES –
AND I'D LIKE TO DISCUSS IT
WITH YOU OVER DINNER –

---

was a brief fashion for walking about sounding like Walter Gabriel, with daisies down your trousers.

Frank Sinatra was a great source of comfort. He made it clear in song that he wasn't getting any either. We used to wear little pork-pie hats and belted mackintoshes and wander lonely as that sad man whose only friend was a Strand cigarette. We'd sit gloomily content under the moon crooning *In The Wee, Small Hours of the Morning; When Your Lover Has Gone; No One Ever Tells What It's Like to Love and Lose; Guess I'll Have to Change My Plan,* and the rest. Now that he is older and fatter, it is clear that he was lying in his teeth. Far from leading the life of merry self-pity he was in fact doing it His Way, being a Rover, and Tasting the Wine in no uncertain manner. He betrayed a generation. Given that and his fondness for Spiro Agnew and brutish men with bulging jackets, it will come as no surprise to you to learn that the tiny hat has been destroyed, the mac burned and that I now identify with Al Bowley, who harmed nobody.

The business of sex

And my word it can be a business, but set to with a will. Sharpen up your foreplay. Oral sex can be fun. Actually, I wrote that down to see how it looked. It looks fine. You probably wouldn't want it emblazoned on a vest, but it's worth bearing in mind. The hard part, of course, is knowing whether they're coming or going, been and come again, or have had a sudden attack of claustrophobia or asthma.

There are those who when asked 'Had a good orgasm yet,

100

dear?' go all bashful, and it's a damned silly time to blush.

There's an old Spanish proverb – *Condicion es de mujeres despreciar lo que las dieres, y marir pas lo que las niegues* – which translated means: 'It is proper for women to despise what you give them and die for what you refuse them'.

There are amazing books to be read. Ignore the *Kama Sutra*, despite the fact it's translated by Richard Burton. It leads you up ludicrous garden paths. Apart from the fact that if two pages stick together you can break a leg.

At the other end of the scale are detailed American books like *The Sensuous Ram* or *The Seven Pullers of Wisdom*. These in the main are discouraging works, telling as they do of the breathtaking potency of the author (invariably anonymous) compared to which the antics of the sperm whale, and indeed the sperm whale itself, pale so that Captain Ahab is tempted to ride again. They contain cheery chapters like 'Getting It Up and Keeping It Up', 'Simple Knots To Tie It In', 'How to Grab Chicks In Museums', 'Catting Around'. Oh, all right, go on, buy one, it may sort out one or two queries, and it's another useful thing to leave lying about her, while you're out there basting.

Radio and television are at it now in an advisory capacity. I watched a fascinating programme on the telly, answering sexual problems. Fascinating in that given television is a visual medium, no one moved. They just sat and discussed. This is no way to set about it. Next week – 'The Gooseberry Bush? Fact or Fiction'.

You can never be too well-informed, but practise, that's where the fun lies. Study time and motion, take your time and move

gently. Pleasure her, pleasure yourself. Let *your* fingers do the walking. It's all imagination, sensitivity, not being afraid to ask, and not shouting 'Wham! Bam! Thank you, Ma'am' at the moment of truth, although you did remember to say 'thank you'.

It's all tremendous fun, whatever the Festival of Light may say. The Festival of Light is not dark enough.

An ugly moment
She: 'I don't know quite how to put this – but – I am pregnant.'
You: 'Then, my word, my solicitors will be in touch with my vasechtomist in the morning.' *Or,* 'Is there any record of immaculate conception in your family?' *Or,* 'Taxi! The nearest recruiting office of the Foreign Legion.'

Embarrassing moment at the chemist
You may remember in that film *A Kind of Loving* Alan Bates nervously entered the chemist prior to a heavy dalliance with June Ritchie and emerged with a bottle of Lucozade.

Let boldness be your friend. It's the only way. Enter loudly. Ask for a young lady assistant, and, projecting well, demand, 'Forty-eight packets of the Gossamer, my good woman, should do me for the weekend'. If she pauses for a moment, to blush prettily – you glance about the premises and whisper discreetly 'Extra Large'. You will be served in a trice, and emerge as the only person in the place not covered in deep embarrassment or envy. (Do you recall when the Consumers' Association filled them with water and threw them out of upper windows? This method received the Papal Seal of Approval.)

102

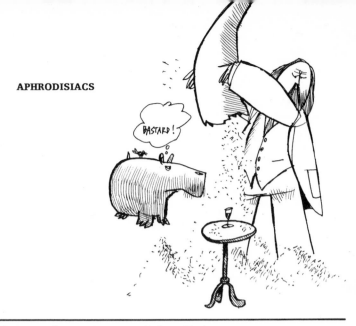

Further embarrassing moment at the clinic

If you think you may be a victim of a dose, repair post-haste to your doctor or if you can't face him, the more anonymous VD clinic.

To avoid the embarrassment of *(a)* appearing to be a patient or *(b)* being recognised by several friends, a simple precaution is to put on a white coat and sport a stethoscope. If you are a victim, you will be forbidden alcohol, so announce to knowing, winking friends that you have jaundice.

Aphrodisiacs

A wise man said that there is only one guaranteed turn-on and that is six months in Nova Scotia without a woman. But the moment you saw the word 'Aphrodisiacs' you were thinking to yourself 'Not me, you fool, a dozen oysters and gin or three, and Dr Jekyll must get his locum in'. What you have in mind is surreptitiously tipping a rhinohorn into her medium dry sherry and repairing immediately to bed with a bottle of pink champagne, a packet of crisps, and a hot lady pulsating like the San Francisco Earthquake. (From which, incidentally, Caruso emerges with very little credit to my mind.)

Well, if all else has failed, try this menu for the Last Supper at Rorke's Drift. (If this fails then examine yourself more closely. Are you very boring? Do you suffer from halitosis? Is your beard alive with tiny creatures? Are you ridiculously un-rich?)

Pickled herrings, liquorice and asparagus.

Avocado pears liberally covered in sesame and coriander seeds.

Oysters with spinach and watercress.

Liver, onions and celery.

All this washed down with a loud Burgundy. Other than that I can't really help you. The balls are in your court.

Suffice to say that many look towards the Mystic Orient for the answer. Actually the answer is 'No', but that wasn't the question.

(Tell me, Mrs Fu-Manchu
Is the rumour really true?
Is it North or South like all the rest?
Or Inscrutable East to Decadent West?)

I read recently that a sales rep. for a Scottish soup purveyors (not to be confused with Scotch, which is referred to on occasion as 'Electric Soup') was demonstrating a can of game soup to a gang of Chinamen. They showed little interest in the contents, but became untypically enthusiastic about the stag at bay portrayed on the label, and in those high-pitched sing-song voices that make Hong Kong sound like a harpsichord taking off from Heathrow, sang out, 'This is what we want'.

With the result that instead of flooding the Orient with thick, rich game soup, these honest Highlanders are now exporting antlers by the thousands. Not only antlers, for when the Chinese discovered further parts not employed in the manufacture of the soup, but a matter of some pride to the stag, they ordered those as well. Apparently, they mash into a jarful of joy. So Go East, Young Man, and don't come to me with your problems. Not that I will ever hear the term 'stag night' again without a sharp pain in the upper trouser.

PARTY-HURLING

One way to lure a lady, or ladies, to your eyrie is to build a party round them. It gives you a good look at them too and in these confusing times separates the men from the women. Otherwise, use a bucket of cold water. Admittedly there's no point in giving parties if you don't like giving parties. There are those who *hate* giving parties, but never seem to stop giving parties. You press the bell and through the letter-box hear a strangled squawk and cry of 'Oh, God, not more *people*' – hosts who spend the evening in the kitchen drinking privately.

There's little to be said for giving parties if you don't like *people*. Look at it this way, you don't *have* to give a party. No one ever pointed the finger at anyone else and said, 'He never throws parties'.

104

If you feel socially obliged to return some hospitality – some couple threw a binge – ask them round for a drink, or take them to the pub one day, if you like them. If not, ignore them, let them cut you off their list, or move. Remember you're British, and it's part of our heritage that we need never say any more to anyone than a civil 'Good day'. How come, blessed as you are with this God-given constituent, you find yourself with social obligations? All right, I admit it, you do find yourself with them. A very good idea, particularly brilliant if you're also cursed with some good boring family or in-laws as well, is to ask them all at once, show blue films and have a nude Zambian transvestite leap from a large sponge-cake. On the other hand, they might like it. Ignore them.

Rules for Party-Hurling

Ask people you like. They will probably like each other – though be it on your own head if you ask all your old girlfriends together.

Friday or Saturday night is favourite.

Friday *and* Saturday night is magnificent.

Rather than asking them to bring a bottle, if you're skint, this is – charge admission, say £1 a head and go out and buy some Scotch or whatever. This way you don't end up with bottles of Riesling (53p) and a hangover that feels like the work of a part-time alcoholic lobotomist.

Food. Forget exotic little tit-bits, they've not come to see you for your well-known cuisine. Good hunks of French bread and British cheese. Smith's crisps and ham rolls. Or get some

PENTONVILLE IS VERY DECENTLY TRYING TO RE-CYCLE A LARGE PINK GIN

take-away samosas or tandoori chicken from an Indian, or spare ribs or spring rolls from a Chinaman.

If they're friends of yours, plastic cups and paper plates will suffice.

Good music. Take up the carpet if necessary. Even rowdier – a piano and someone who can play and enjoys it. I know it sounds old fashioned but you can't beat a musical evening of the old school.

Drink. If it's not a bottle party or a pay-as-you-enter, keep it simple. In winter, Scotch, beer and large flagons of Italian Red cover most eventualities. Water is all you need with the Scotch, ice for some, ginger ale perhaps as a sop to several.

For a touch of the exotic, try something different in the beer line, like Australian or Polish – being more lager-ish, the ladies may be more keen to plump for them.

Forget gin, you need too many mixers. Neat vodka, however, from the fridge, thrown down in one, via an egg cup, is a good recipe for a very short party.

In summer – cold, white wine – many victuallers pack two-litre bottles of Frascati, and it goes down very easily. Lager again is most welcome.

Whatever you have, always have enough. If you're getting it all from a wine shop, they'll probably let you have it sale or return, and let you have some glasses if you're that way inclined.

Smokers. Ash trays galore, strategically placed. Hide away spare packets of fags, thus avoiding that dampening moment when they run out later.

Warn the *neighbours* nicely, and if not officially inviting them, ask them to 'pop in if they want to' lightly. Then if they come to complain, slap a glass in their hand, and start introducing them briskly. 'This is Arnold, our good neighbour, thought it was a pyjama party.'

Tell the spirit-drinkers where it is and they can help themselves. The wine and beer can be passed ceremoniously from group to group. It encourages light banter.

Gate-crashers. This only happens if you're sharing a party and don't know who the other has invited and vice versa. If gate-crasher is established: *(a)* Make him/her hideously welcome. Silence the assembly and call upon him for a speech. *(b)* Prime several to approach him/her and say loudly 'And who the hell are you?' This I have seen work very well. *(c)* Get someone large to throw them out.

The guest who will not leave: You've tried circling him slowly in your night-shirt, yawning ostentatiously, and saying 'No, no, you're not keeping me up'. You've remarked several times on the excellent firm of mini-cabs whose number is by the telephone. Try this. Be seen to be preparing an extremely uncomfortable bed either on the sofa, if you have one, or the floor among the debris. Open the window saying, 'Sorry, there's only one blanket'. Apologise for lack of pillows, but triumphantly wave old cushion, sniff it suspiciously and say, 'Bloody cat'.

Invent a dog if you haven't got one, and warn guest not to be alarmed if it attacks him at about 5.30 in the morning.

Leave guest saying, 'Popping off now. Got to get up early – so may not see you. There's no coffee, but there's an old tin of prune juice and I think the bacon's all-right but I haven't seen it recently.

Redraw his attention to the mini-cab firm's number.

En passant for a party In Australia people have a pleasing habit which is, when dropping in unexpectedly at any time, to arrive with a bottle. It's part-bribe, of course, but I'm anybody's for a small, convivial gathering.

WHICH SIDE DO YOU DRESS ON?

The outside, you goat! But clothing, though more casual nowadays, presents more of a problem than in days of yore. The days of yore basic uniforms of dark suits for the office, yore cavalry twills for the weekend and yore duffle-coats if you were mildly avant-garde. They're jollier now and infinitely more comfortable and expensive, but they betray you at a glance,
108

The Hooray Harry – in chunky sweater, wide-checked Viyella shirt, Paisley or Old School scarf, tight twills and suede bootees.

I once offended a whole gang of these sporting heroes of the 3rd XV by saying that Rugby Football could only be discussed properly in Welsh.

especially if you are trying to impress. One false turn-up and you stand condemned as *very* Third Week in August, 1976.

Suits

The cheapest suit I ever worked out was found by going to some vast emporium where I bought separately a brown corduroy jacket and a pair of matching trousers. Indeed the same in black cost me about £9 and served me for many years as a dinner-jacket.

A dinner-jacket is a waste of time and money, unless you're a waiter or gigolo. I find a brown velvet suit serves with a tasty white shirt and a made-up brown bow. A blue one would also suffice and you can wear it at less formal functions, even the office.

The answer would seem to be to have one suit that is definitely office, but preferably with two pairs of trousers, so that you can juggle with the dry-cleaner.

I am a great believer in waistcoats. They mean that you need iron only the collar and cuffs of your shirt, and they give you valuable pocket-space.

Whatever happened to pockets?

A disappearing breed. You're lucky now if a trouser has any at all. *To repair holes in pockets* – use a stapling machine.

God knows, I'm no Beau Brummel. I think quite a number of people are aware of the fact. But whereas the peacock has but one I have three suits, including the brown velvet I got in a sale. Given those, and one pair of brown checked trousers, it's

109

The Man who has Conquered the Fast Lane of the Reigate By-Pass single handed. These jackets seem to be all the rage – they're warm enough but leave the bum exposed to the elements, which is fairly silly of them. And those strange back-less gloves – what are they hiding? Hairy palms?

amazing the changes you can ring. And the odd whistle you can raise.

One is a light cream, the other brown and woolly, but I can, for instance, wear cream waistcoat and trousers with dark brown velvet jacket and evoke memories of the Edwardian era just like that. Or wear the checked trousers with the top half of the brown and woolly and look still dated, but quite refined. What I am armed against is the sudden exploding zip. The few clothes I possess do have reinforcements and are readily inter-changeable.

The sudden exploding zip.

There's not much you can do about this except say 'Hello, Erica Jong – here stands a zipless fuck!' You can dive into a shop and buy another pair of trousers or if all else fails, as a purely temporary measure, bind together with a length of black insulating tape. Find a dry cleaner who does repairs. Which reminds me that in Australia sticky tape is known as Durex. It's not nearly as efficient and bloody painful.

Once you're suited up and have sufficient to get you through the body of the week, what you buy else depends on your taste and budget. I *would* recommend that you get out of your suit on returning home, however worn out or legless, *(a)* it makes you feel better and *(b)* if you hang it up, it may make it through tomorrow.

Casual clothing is a personal matter though. And a good thing too. A simple test. Identify with one of these characters and if you score over 57 per cent, forget it.

110

The Lone Ranger or Macclesfield's Own Midnight Cowboy. Probably an advertising executive – his T-shirt will tell you. This is a popular line, but again useless in cold weather. You have to keep polishing the leather hat.

Hats

I am a firm believer in hats. They should, however, be hats of character. A good large hat *(a)* keeps the rain off and is less burden than an umbrella, *(b)* disguises the fact that you've mislaid your comb, *(c)* is a very useful receptacle, for instance in the boozer if you have odd bits and pieces you can't pocket – put them in your hat along with your head, or if they're bulky, remove head and pop them in. An inveterate hat-wearer never forgets his titfer. I keep messages to myself in mine.

Scarves

Again a useful piece of apparel. A long woollen scarf *(a)* is a very useful emergency windscreen cleaner, *(b) can* be used to lash down a bursting suitcase, and *(c)* keeps your neck warm.

Top-coat

There's very little point in one unless it's waterproof. The average old-fashioned over-coat is a boring object. You see a lot of fur coats nowadays on men. But a light monsoon and the smell and appearance are appalling. A good pseudo fur-lined car-coat is quite sensible and I also like those monstrous camping macs with furry hoods and pockets in the sleeves for cigarettes. They do keep the weather out, which is, after all, the point.

Shoes

I've always hated laces, and elastic sides seem to give out. I tend towards boots of various lengths. Again the more sturdy the better. If your feet are warm, you're warm. In winter I wear old football socks with them, you can pull them up over your

This is as casual as some people get. You find them at the Races with their funny little brown trilbies and shaggy coats or British Warms. Green tweed seems to be the thing. They'll never change. They should be constantly harried by the Noise Abatement Society.

knees. Conversely, if your feet are cool, you are. So in summer wear, when possible, those flip-flops that you keep on with dextrous movements of the big toe.

The other advantage of boots or bootees is that in emergencies you needn't wear socks at all and no one will notice.

Smelly feet These are highly unimpressive. So fire foot-powder down the boots, don't wear nylon socks and wash all feet frequently (in the bidet's best and avoids groin-strain in the basin). Use pHisohex available at all chemists. It's a liquid soap as used in hospitals – and works harder than soap.

Underwear

Never too tight – the underpants. It leads to all sorts of unpleasantness, including Dhobi's Itch. Particularly nylon which causes sweat rash. Cary Grant, I once read, wears ladies silk underwear, and it seems to have done him no harm.

As to a vest – wear T-shirts. You are then ready for the summer, especially if you have dyed all your underwear some exotic colour (See page 34).

Undressing

Never forget the old words of wisdom about how your underwear looks – you never know when you are going to get run over unexpectedly by a bus – and all will be revealed. Nor do you know when you are going to reveal all unexpectedly to a lady – never allow yourself to be inhibited by the knowledge of your tatty pair of knickers – keep them in order; you never know when your moment will come. When it does don't forget to take your shoes *and* socks off first.

112

THE PARAPHERNALIA OF LIFE

This section concerns kids, household pets, plagues of locusts and perfumed gardens. The only justification for lumping them all together is that each in its various ways concerns the balance of nature – that is to say, you give birth to kids, raise them, nurture them, feed them, care for them, work for them, and the effort finally kills you. That is the balance of nature.

When you look at with what you've been lumbered
It's no small wonder your days are numbered.

KIDS

As a single man, you are probably either a full-time mother and father or a part-time father with reasonable access. (No relation to Unreasonable Access who are a credit card firm.) Chances are, therefore, that your youngest is five or six and your eldest could well be 21 or more and older than you are. It's reasonable in that case to do large diagrams of how to change nappies. (Though, my word, I used to be proud of my nappy-changing. Nowadays of course they have brilliant instant ones, and the craft is dying.) Who am I to tread where Dr. Spock has boldly gone? Essential reading, anyway, you'll find a few pointers towards dealing with the 21-year-old or over. I hope you were involved during those first five years, giving suck on occasion and the like. It makes it so much easier later.

I remember when Toby and I first met at St George's Hospital, and I held the tiny gobbet in my large arms. 'Kiss him' they

113

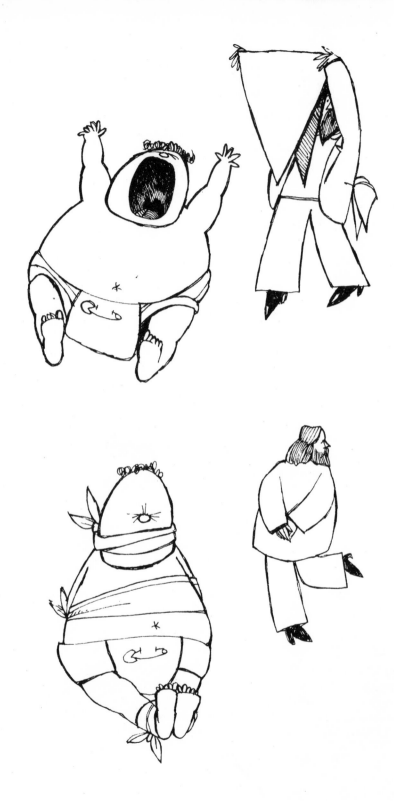

cried. The Englishman in me reared up at once. 'I hardly think
. . . What? Good Lord . . . I'm his father . . . He a son . . .' I saw
all those ladies' faces tauten, and to advantage, I've kissed him
vigorously ever since. Apart from the pleasure, it keeps my
beard in check. Holding hands is a constant source of delight
too. And hugging. Think European, we might as well get something back for our membership, the annual subscription is
appalling.

There's a lovely line in *A Thousand Clowns* which Jason
Robards delivered in the movie version. It went, roughly,
referring to his young nephew – 'I would like to tell him the
simple, sweet reason why he was born a human being, and not
a chair'.

If you work on the understanding that kids are human beings,
more human than you, and in no way designed for sitting on,
there is no reason why you should ever have any problems with
them, even if you live with them.

Bags of warmth and affection, love and adventure and
laughter. Add a dash of spice and excitement to everything you
do. When driving to Granny's, make sure you get lost on the
way. If possible in a forest, with a fair chance of being
enchanted. If you must have a haircut, have them together. If
she's a little girl, a ladies' hairdresser is even better, lads.
Enjoy that hour under the drier with Marjorie Proops. (You can
learn a good deal from women's magazines about their tactics
and stratagems.)

The joy of kids is that they stretch you (you always thought
you needed eight hours' sleep), they drag you out into the open,
flavour your days, make you laugh, and make the tears flow from
ducts you thought long extinct. They're a charge to the heart, a
sock in the parts, and, oh, the constant challenge every time they
fix you with those wide eyes and put the simple question 'Why?'

It's a moment not dissimilar to the one when you're faced
with a sudden lion, the accepted rule in that case being to stare
pointedly back and not to budge. Lion-tamers seem able to
command the same respect with a chair. This can also, in
emergencies, work with children, but I tend towards non-
violence, identifying more with Destry than Wild Bill Hickock.
(Admittedly, Destry wound up with Marlene Dietrech but into
every life a little rain must fall.)

Rule 1 Answering
When the question 'Why?' is put to you. Answer it frankly.

DON'T YOU TAKE YOUR FIRST STEP TOWARDS ME, RAMPAGING HOOLIGAN!

You may find for the first time in your life you're contemplating the answer 'Why not?' To say 'Because I say so' is avoiding the issue and is also the sort of remark which started World War II.

Rule 2 Shouting

This, used sparingly and briefly, and only when you know you're right, is O.K. when it's all gone too far, and there are those occasions. However, my small son, when he knows that *he* is not the actual object of my raving, but in fact, it's the world at large, or the Inland Revenue, or the bus in front, is allowed to say gently 'Come off it, dad'. If he's right, and I'm wrong, I apologise. And he's really very nice about it, and will sometimes join in. That's what the father/son relationship is all about. Shouting in unison at the world at large, or the Inland Revenue or the bus in front.

Still thinking European when you look abroad you realise what an uptight isle Britain is. There's nothing more delightful to watch than the French *en famille* in some restaurant at a table for 22, ages ranging from two to great-grandfather. The brouhaha is enormous but nobody cares a tittle. The which gives rise to Rule 3.

Rule 3 Panicking

a Do not allow yourself to be cow-poked, that's not right, bull-dozed (slightly better) into embarrassment or alarm by, for instance, caustic glances from older women in restaurants, who make it clear with a nostril that they think you should be locked away by the NSPCC for allowing your child to sit under the

116

I SHALL NOT PANIC - THESE ARE NICE OLD LADIES

table amputating Action Man's right leg with a spoon. Stare back (*See* Lions – how to cope with manfully). *Or,* and I tried this successfully on one occasion and Toby enjoyed it tremendously, sending him over after a brief rehearsal to whisper, 'Here's looking at *you*, Bright-eyes' in her blue-rinsed ear.

b If you are, so to speak, entertaining or even doing business in the home and are approached by progeny, who only want to see what's afoot and remind you of their existence, do not send them away with the proverbial flea. Introduce him/her to him/her, take a couple of minutes off to explain what you're at, that it's boring old adult stuff and that it will take another five minutes, half an hour, or whatever. Don't be hustled by any signs of adult impatience.

c Never be afraid of silence. Take a quiet gander, but in the main, silent kids are contented kids. You'll know when they're bored.

I remember reassuring the cast of *Gulliver's Travels* which we did for kids at the Mermaid Theatre around Christmas 1975 when they worried about the silence.

After one show some 10-year-olds were saying how much they'd enjoyed it.

'You were strangely quiet', I said.

'I laughed out loud once', replied one, 'and Teacher told me to shut up'.

Proof positive that adults panic needlessly. Give them their heads, they're older than you think, on those young shoulders.

d There is very little point as a guest leaves in bellowing 'Say "Goodbye" to the nice person who is leaving!' The

Forget it — we are lumbered —
there is no avoiding Granny's Hot Pot

Kiss Auntie at once — regardless
of Haliotis and bristling moustaches

Tell the nice Customs Officer we
cannot wait as you are going
to be violently sick.

'Goodbye' if you get it means nothing and further silence can leave a neurotic guest with a permanent identity crisis. Mouthe it wordlessly if you must.

Cultivate a language of signs. A father/kid conspiracy can be quite fun in their eyes. It works well in the case of an ageing aunt. 'Ageing aunts', you can say, 'are like many adults and can be quite silly, so we must be kind and nice to them and watch our language and not set fire to their handbags'. Suggest a Be Kind to Auntie Day. We quite often celebrate a Be Kind to Dad Day. 'Anyway', you can add, 'Auntie invariably comes equipped with a bribe. So don't set fire to her or use her teeth to remove Action Man's other leg.'

Bribery

Not a good idea. Buying a kid's silence for 10p is non-productive in the end, and could lead to a life of crime at worst and inflation at least. Sweets for doing this and chocolates for doing that smacks slightly of circus-training. I think in cases of visits to the dentist and that sort of event, a stop-off on the way at a toy emporium, but with a limit, say 30p and that's it. Money does not grow in England, it needs oily sand, this is acceptable.

The odd bargain can be struck. You must be allowed occasional self-indulgence. (Another good thing about having kids around is that it makes it very hard to be selfish long, and disaster looms if you are.) It's fair to say on a Saturday night: 'Look, it's half-past ten. I watched *Dr Who* and *The Parade of the Zombies,* didn't interrupt once, made no noise at all, and fed you under the bed, so now either retire or sit quietly while Dad watches Vanda Bosom do something salacious that is bound to upset Mrs Whitehouse, but will bore you rigid.'

It is certainly disgraceful behaviour to pour expensive gifts on your lady-friend's kids in the vague hope of seeming indispensable as a pseudo father-figure. Or indeed to offer advice on how to handle them, or taking their side against her. If her kid is slowly clogging your hair with modelling clay, and mother seems not to care unduly, take them quietly outside and threaten legal action. On the whole, lie back, relax until they accept you.

Table Manners

This gives rise to panic in some. Conspire. Have different grades of table manners. Hotel restaurant manners or table-cloth manners for instance. Make the meal as quick as possible. Do without your coffee and port, forget a starter, just have a main course and let the kids dwell on the puddings to come. Point out though that it's hotel restaurant time and certain rules must be obeyed, which is boring and proper. Observe that father is on his best behaviour, wearing a tie, and trying to devour soup silently and not singing between courses. Not a bit like home.

a Don't put your elbows on the table! Why not? Good question. O.K. Put your elbows on the table if you can be sure not to knock that glass over.

b I long ago gave up the traditional British method of pea-eating. Why? Why not? I certainly can't explain it to the younger generation, nor will I bother.

c It is a good moment to encourage the 'napkins not sleeves' technique. Now that is practical. The Done Thing is quite often the silliest thing to do.

At home, keep meals cheerful. Let them go when they've finished. Don't force food down them. 'One for Uncle Arthur' may sound like a reasonable swap, but how does Uncle Arthur feel when his one whistles past his left ear?

If they want to eat on the floor, watching the telly, simply put newspaper under them.

FEEDING THE KIDS

They'll eat what you eat, and probably prefer it to a constant diet of fish fingers, hamburgers and chips and boiled eggs and baked beans. It's good to have a chicken or a roast or even

119

IT KEEPS THEM QUIET FOR HOURS

Psychedelic Spaghetti together with them (see page 61). Then again, ring the changes and play their requests. If you get a sudden demand for chocolate ants there's probably just been a commercial for them on the box. But a cry of 'spaghetti hoops!' should not be left unanswered. (Mix small squares of buttered toast in with them. It adds a certain verve.) Occasional rubbish does no harm, particularly if you drop a few vitamins, like Adexolin, into them with their morning tea.

Puddings are simple and should be, otherwise you'll find yourself slaving over a Bombe Surprise nightly. Yoghurts are popular. Apples and bananas. Cheese and biscuits. All good stuff. And there's always ice cream. Vanilla and hot chocolate sauce is much fancied.

Tea time
Young and old can go potty together in the kitchen around 'le five o'clock'.

Making pancakes enlivens a dull day.

Pancakes
$\frac{1}{4}$ *lb flour (plain or self-raising)*   *1 egg*
large pinch salt $\frac{1}{2}$ *pint milk*

(Add 1 tbsp of melted butter if you like)
Sift the flour and salt into a bowl. Add the egg and half the milk (butter too if you're including it) and beat well either by hand whisk or blender. Then stir in the rest of the milk. (There's a job for everybody so far. Sifting, beating eggs, stirring.)

You'd better handle the pan (iron is best) which should now be
120

hot and sufficiently greased to stop sticking. Nothing is less impressive at the moment of tossing, which can be the high spot of the day, than a pancake which refuses to budge.

Have different fillings to add variety. Eat them as you go.

In the summer you can have a *pancake picnic.* A packet of pancake mix, a plastic bottle of water, a camping stove (Gaz or whatever) and your pan are all you need. Serve the pancakes on paper towels.

Waffles

It's well worth buying a waffle-iron. If you get one it'll probably have the instructions with it but in case you've borrowed one:

8 oz flour	*3½ oz melted butter*
5½ oz sugar	*pinch salt*
3 eggs	*¾ pint cold milk*

Put everything except the butter in a bowl and beat thoroughly. Then add the melted butter.

The waffle-iron should be very hot and constantly re-greased with a nice buttery pastry brush. Fun for the artistic child. Don't put too much mix on at a time or it squirts out of the sides and attacks the cooker.

Slap on ice cream and maple syrup while they're hot. Listen to the tiny cries of 'Yaroo' and 'What a father!'

Cake mixes

Further entertainment. Buy one of those 'Just-add-an-egg' type. The ones for small cakes provide their own paper cases – so there's no washing up. But a sense of triumph. I remember Toby and I had to create a cake for the school jumble. We bought three chocolate cakes of differing sizes. Mounted them one upon the other, welded them together in the oven, and decorated the top with smiling face composed of Jaffa Cakes and Smarties.

It's a good idea if cake-making becomes a popular pursuit to keep a supply of decorations, chocolate strands, hundreds and thousands (be wary!), silver balls and glacé cherries.

Jam tarts

Have you ever seen one refused?

Put the oven on at 450°F (230°C or Gas No. 8). Roll out de-frosted pastry quite thinly. Don't stretch it. Cut out rounds of pastry with an upturned glass or plastic cutter and put in tart tins (non-stick if possible). Brush them over with milk so that

121

they look shiny when cooked, if you want to, in case they fail to stand comparison with the life-work of McVitie and his friend Price, or Mr Kipling. Put jam in each. Put them in the oven for 15 minutes. For swift cooling, as you will be surrounded by a drooling mob, wave them out of the window or fan with your hat.

Icing sugar
Easy and it tarts up a boring sponge if that's all the shop had left. (It also sponges up a boring tart, but not in front of the children.)

8 oz icing sugar

2 tbsp warm water

Sieve the icing sugar into a bowl. Add the water slowly and mix well with a wooden spoon.

You can add colouring now if you like. Go on, go mad. *Only a drop or two at a time* or it may look as though you've bled on the gateau.

If it's too runny, just add more icing sugar. Spread over cake or lots of fairy cakes. What fun you can have, and the children can have a go if you'll let them with one of those icing bags that squirt out decorations and friendly messages. This requires practice. Start *now!*

Peppermint creams
This is a good one, if you're surrounded by very small kids and are slightly alarmed at the prospect of hot pancake descending on their innocent little heads.

122

8 oz icing sugar $\frac{1}{4}$ *tsp peppermint essence*
1 egg white

Sieve the icing sugar. Put the egg white into a bowl and whip with a fork until light and frothy. Give the kids a go. Then slowly add the icing sugar, working each bit in before adding more. When it's getting stiff, get your hands in and wrestle with it. Now sprinkle some icing sugar on working surface. Removing from bowl, slap down the mix adding the peppermint essence. With ferocious noises, always good for a laugh, knead like fury. If it's a touch soggy and damp, add more icing sugar.

If just right, roll out into $\frac{1}{4}$ inch thick slab and cut into any shapes you like. You can buy animal-shaped cutters quite reasonably. Also good for a chuckle. 'Here, Norbert, is your praying mantis-shaped peppermint.'

If these afternoons of culinary delight become popular, and they're a fine example of the pleasures of doing things together with jobs for everybody, you'll even find the odd kid who feels important washing up. Keep a constant weather-eye and all will be well. Get too many ingredients there's bound to be a minor calamity and it proves you're as incompetent as they. If they're popular, there are some excellent cook-books for kids. Peruse the bibliography on page 181.

Barbecues

Just a thought, *en passant,* if it's a nice day. If you've got a barbecue you've already thought about it, if not, build one, it's an adventure. Very well, you're like me, and we're all right because Woolworth's sell disposable ones, made of foil containers with a wire grid to go over the charcoal. You can take this on a picnic and cook hamburgers, sausages or chops and have baked beans or buttered buns with them, all washed down with cans of Coke.

First aid

This seems as good a moment to mention it as any – and is no reflection on Woolworth's disposable barbecues. Never be embarrassed to call the doctor or if a kid's had a nasty fall or similar thump and something might have fractured, whip him to the nearest casualty department as quickly as possible. The same applies if all seems unbroken but he starts passing out or appears particularly dozy after a tumble.

Cuts and Grazes Get all the dirt out. Soap and water will do for the minor ones, but a drop of disinfectant in the water is never a bad idea, and the sting makes the kid feel heroic. So does a bandage or sticking plaster, even if unnecessary. Savlon is a good soothing ointment. Keep a first aid kit in the car.

Minor burns and scalds It's a good plan to keep some impregnated gauze about for these occasions. Otherwise Savlon or somesuch with a clean dressing will suffice. I still stand by a squirt or two from a Vitamin E capsule, but if it's a bad burn get the doctor.

Pinched fingers Fairly frequent, this one. Over-excitement around doors, etc. Stick the finger under the cold tap to prevent blood bruises. If the poor kid's in a state of mild shock, give him/her a hot sweet drink or something to suck.

Wasp stings or similar Highly popular in the summer. Keep some Wasp-Eeze about. It neutralises the sting at once and prevents swelling and pain.

In general, ooze confidence and sympathy. If the kid is having a good cry, let him/her, it's better out than in, that's why we're given that ability. Only become testy if it's a con-trick.

I wonder if the Englishman is what he is because he was told not to cry when small. After all, if you can control tears, stiffen the upper lip in an instant, then you'll probably under-exercise the other finer feelings. I've heard it said we were born two large Scotch under par. I think it comes later. Two large Scotch, please.

HOW TO OCCUPY THEIR TINY MINDS

This is not a 'cri-de-coeur' because it hasn't got a question mark. It can be a problem though, and a lot obviously depends on the age of the children, where you live and what's going locally. Check around, and you might be amazed at the facilities available. For instance, *(a)* for under-fives, see if there are any 'One o'clock Clubs', etc. *(b)* For the older ones, check out the adventure playgrounds and such like.

But, of course, a further deal depends on what sort of father you are, part-time or permanent.

If you're part-time, plan out, in committee, what you'll do together next Saturday or Sunday or whenever. This gives *them* something to look forward to, and they're all excited when you collect, and you don't fall into that fearful trap of having to 'bribe' them to like you every visit. They'll suss you out, sir, in a nonce. They'd rather have your time, energy and madness than a banker's order. Security not securities. Amaze yourself, while amazing them. It's second childhood time. Who thought you'd ever see the Tower of London again? Or build a sandcastle? Or

Box labels:
- God Knows
- Remains of Books
- Weapons. The Rest of the Lego. Odd socks. Cars (with wheels). Cars (without " "). Wheels (without Cars). Burst Balloons
- Works of Art. Minor Masterpieces. Further Lego. Plasticine. Chewing Gum
- Natural History. Dead Frogs etc.
- Dinosaurs. Lego. Half-Eaten Toffees
- ACTION MAN. Surplus Limbs, Heads, More Lego
- WHERE DO WE FILE AUNTIE'S DENTURES?

ring a doorbell and run like hell? (It gets a grip of you, doesn't it?)

If you're permanent, it's a different bag in that kid or kids will have own room and a fine old chaos-pit that doubtless is. Now there's chaos and chaos, and if you look in a kid's room and scream 'Chaos!' and he looks in yours and shakes his/her head sadly and you say 'Ah, but this is well-ordered chaos', you're on a loser.

Suggest firmly that as a family it's high time all pulled together, and so all together it's bullshit time, or words to that effect.

Kids don't mind a bit of hoovering, and wandering about squirting patent sprays and waving J-Cloths and feather dusters. It's all part of the adventure of living together. Having all rallied, then it's time for *all* to celebrate. *We* all deserve a treat. They may not think that sitting outside the pub for an hour is any sort of treat. Compromise.

Elementary bullshit

Boxes Cardboard boxes from the supermarket, or fancier ones if you like, open at the top so that you can sit in the middle of the chaos accurately hurling Lego in one direction, Action Man's necessaries in another, doll's removable eyeballs left, stuffed pandas right. It makes cleaning up much simpler and improves your fielding and darts. Admittedly as you hurl the ball yards over the wicket-keeper's head it is a feeble excuse to say: 'Had that been Action Man's leg, it would have been precisely the required three inches above the stumps'.

125

WHEN I GROW UP, I'M GOING TO BE A MULTI-% MILLIONAIRE

You can get those splendid *double-bunks with drawers underneath,* and thus, in a very small space, you can file away two kids and a mess of trouble.

Bookshelves can hide a multitude of objects easily, even books. Make certain they can't fall down.

What to do however

Part-time or permanent, there's still the business of finding areas of mutual enjoyment and entertainment. It can't please everybody, of course, but if parent's worried if kid's enjoying self and vice versa, no one's enjoying anything.

Never ignore the old hobbies – *stamp collecting and scrap-books* and the like – always keep masses of paper about, lining paper is a good notion, and paints and crayons and felt-tips, etc. That's how Picasso started.

Junk. Keep a box full of old cereal packets, lavatory rolls, milk bottle tops, washing up liquid containers. These can be turned into anything. There are a number of books, covering this area of juvenile do-it-yourself – see them listed at the back of this book.

Use your local *public library.* Apart from being a useful source of books on fun things to do – they also post local events for kids and sort of amateur Jackanories on Saturday mornings in the holidays, which gives you a free crack at the grocer.

Also keep an eye on your *local museums.* If you are in London you're lucky with the Science Museum and the Natural History Museum, both of which have film shows and lectures, not all that fascinating, but ones about space or dinosaurs can't be bad.

The *local paper* too will tell you of other likely events like air displays or open days at the fire station, fêtes and jamborees and visiting circuses or wandering bands of Thespians, catering for kids.

Dear old Britain is thick with *history*, it lurks round every corner, but it's invariably so boringly taught, that you'll be doing the kids a favour if you whip up some enthusiasm by visiting the houses and castles they've heard of, but, in a book, couldn't care less about. For approx. £1.50 you can get an annual season ticket (kids 75p) to all ancient monuments and historic buildings run by the Ministry for the Environment.

En passant, come a birthday or similar celebration, rather than go berserk at home, invite a manageable number of your kid's small friends and go on an outing.

London Zoo I know allow even the most un-zoological of us to become Associate Fellows which sounds quite grand, and for a modest fee you have free access for a year to the zoo and the use of a very comfortable club house, who will organise a tea-party for kids if suitably warned. The beef sandwich is a beef sandwich, there is no need to enquire what beasts have died recently – zebras, etc. Apart from that, expensive they may be, some-such safari park or stately fun palace as Woburn Abbey provides an afternoon of high adventure. The children may wonder in the case of the Abbey what strange order of holy man believed in travelling by cable-car over fields of giraffe, but if they wonder too loudly, suggest a game of Trappists.

Apart from the obvious *sports* like football, tennis and swimming, there are myriad other activities – and it's well worth browsing through the books and brochures published by the English Tourist Board (see page 101). You will not fail to boggle at the things you can get up to, and learn together. Holidays on barges, sailing and pony-trekking, hang-gliding, pot-holing, parachuting, bird watching and hot-air-ballooning. All just waiting for you. Wherever you are there is a local Tourist Information Centre, and it's extraordinary how much more the tourists get to know about where you live than you do.

What about *flying a kite?* Or a model aeroplane, built in your own living room? Watch it disintegrate at a height of four foot three inches.

Go on a *nature ramble.* Collect wild flowers and interesting leaves. Press them into a book and find out what they are.

Collect stones of strange shape on beaches and paint them later. Or polish them.

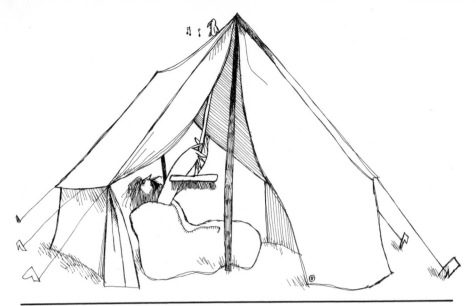

Start a museum of your own with memorabilia of all the places you've visited together.

Go to a holiday camp. The kids love it and they make me laugh.

Go camping. That probably stopped you in your tracks. It costs very little and is jam-packed with adventure.

Any bookseller will have a good camping guide, or a copy of the magazine *Practical Camper*, giving you lists of sites and amenities. Find one near a pub and you're gently chortling.

What sort of tent you get depends on how long you intend to stay in one place. Unless you are a fully paid-up Bedouin, a simple ridge-tent will suffice and probably have a built-in ground sheet. For the fully paid-up Bedouin there are now two-storey tents with spare rooms and balconies and kitchenettes. It's a rare afternoon out with the kids going to a camping exhibition, you can happily lose them for hours in a canvassy maze.

On the Continent, le camping is an infinitely more sophisticated matter, and quite often a deal more comfortable than the local hotel. All the sites seem to lack are bedrooms, hence the tent. They provide restaurants, take-aways, swimming pools, shops, etc. There are two **ways** to make it easy: *(a)* Through a travel agent. There are **packages** that provide everything (save possibly for sleeping bags) **and the** cost of the ferry. *(b)* Townsend Thoresen, the ferry persons, operate a hire service, you collect what you need on board, and if you're going there and back with them, you are blessed with a healthy discount. Unless you have considerable storing-space, hiring would seem the sensible solution, and then it's heigh ho for the open air.

128

WHERE DO YOU STAND?

If you are left holding the baby remember there are lots of others like you and there are a number of organisations that will provide advice, comfortable words and a cup of hot, strong tea. You'll find their addresses at the end of this tome. I've said it before, and I'll say it again – never hesitate to ask.

PETS CORNER

If you are totally alone I suppose it's quite a comfort to have a dumb friend lurking. I've lived with a tortoise until, alas, old age caught up with it one harsh winter. It had lived happily for some years in various flats, loved carpet, but seemed unable to get to grips with hibernation, particularly if thrust into a cardboard box full of Harrods straw as recommended in tortoise guide-books. God knows, I showed it the book, and it still rejected the idea. Nevertheless it makes a change from 'Down, Rover' and 'Sit' to wander round a flat shouting 'Hibernate, sir, hibernate'.

It used to come when called after the day or two you had to allow for the trip. Fetching seemed beyond it – I still find golf balls in the strangest places and think of my late friend, so affectionate, so peaceful and so easy to run. In death, I must confess, the smell was appalling and his shell went mushy. But this is perhaps true of all of us.

It reminded me of a remark of Gerard de Nerval who asked why when walking through the *Bois de Boulogne* he led a lobster behind him on a length of pink ribbon replied, 'It does not bark and knows the secrets of the sea'.

Man's best friend

Ha! If you think you're eliciting Mrs Bradley's telephone number from me, insatiable rogue, piss off!

Man's second best friend

I'm not sure where the idea originates from that dog and man are soul-mates. Quite frequently, it depends on the sex of the dog, but bitches seem to prefer blokes and vice versa.

If you're alone and work at home, then a dog is fine, but if you're in a city on your tod and work all hours, you're consistently stiff with guilt about Rover. In the city, anyway, it strikes me as only fair that you should have a smaller dog,

otherwise exercise is a fearsome bane. (They are also a prey to dog-nappers.)

I used to walk a miniature long-haired dachshund round Notting Hill till a young wag shouted 'Who sat on his dog, then?' I had to laugh, but I never felt the same again about it.

However, basic rule

Fit the pet to your life-style. If you own a safari park it is foolish to dote only on ferrets, however charming. If you live in a bed-sit, be it never so tempting, avoid housing an orang-utan, however much it resembles mother.

If you travel a lot, tortoises and goldfish are survivors, but you'll worry about a dog or a cat. You can always park them with a willing lady. It forms a small but useful bond.

Cats

Now they are of a more independent mien. Ruthless, some. They can be left around the place without your panicking unduly. Particularly if you have two and they can tear the place apart together. A French lady pointed out to me a further advantage of the *chat* as they have it over there, – she said (and these Froggies are well into this sort of thing), that a cat is a sensual creature, and reacts favourably to candlelight and soft music, indeed has almost aphrodisiac qualities, with the purring and the stroking. A dog on the other hand, or on your lap, or suddenly and violently on your upturned naked bum can destroy an ambiance in two wags of its tatty tail.

You can't put them out either.

130

The miniature long-haired dachshund was called Mrs Saunders. I thought that was her name from the birth certificate, it turned out to be the breeder, but it stuck. Thinking it was high time she bore fruit, I sent her off to Bath to said breeder who arranged a moment or two of canine bliss with Mrs Saunders' grandfather. This is apparently *de rigeur* in doggy circles. I regaled a very boring young man with this tale at a party that evening. Unthinking, and perhaps wistfully, I said, 'Mrs Saunders has gone to Bath to be fucked by her grandfather'. 'My aunt', he replied, unmoved, 'has just been to a health farm'.

Naming the beast correctly is vital. I met a man in a boozer who had two fine labradors called Castor and Pollux. Whereas he said he could happily roam the neighbourhood calling for the former, it was an act of madness to walk the streets bellowing the name of the latter.

I leave the choice between cat and dog entirely to you and your circumstances. Much depends whether you have kids or not. A dog can be a spendid companion for an only child, I know. They can talk for hours happily.

Goldfish

Very easy to run and quite attractive and very quiet. We won four once at the Hampstead Heath Fair some years ago. One vanished almost at once down the plug-hole during a clean-out, another played gooseberry for a while until the remaining duo ate it. That was all quite exciting and doesn't happen so often with dogs and tortoises. Nature in the raw in goldfish bowl yet.

They do appear to have remarkable powers of survival. I

bought a house in the country years ago which hadn't been occupied for six years. There in an outside pool was a goldfish happily carrying on regardless. We christened him Portnoy, I remember, and bought some enticing mates for him which he shunned.

Mynah birds

These are good value, incredibly messy and can lead to confusion.

The good value comes from the fact that they are highly entertaining if you can still tolerate impressionists, and happy hours can be spent teaching it simple phrases and bursts of song. They're like feathered tape-recorders. I always liked the American who taught his mynah to drawl, 'Bugger off. Birds can't talk'.

The incredible mess comes from their boundless energy and seemingly limitless incontinence. They leap at a pace from side to side of the cage discharging waste in all directions. You end up with plastic sheeting all around the cage and several editions of *The Times* (quite the best for this) under the explosive bird. In themselves, they are the cleanest of creatures, and bathe frequently.

The confusion arises from the accuracy of their mimicry. This bird could say a rich, fruity 'Good morning' in such tones they were so indistinguishable from my own that even my mother discussed world affairs with it for half-an-hour before realising it had not sprung from her loins, but an egg.

In general

Buy pets from a reputable pet shop, and make certain you get a good book of instructions. And a friendly vet.

Like people, most pets need care and attention, love and the odd kick in the *gluteus maximus.* Don't get a pet if you can't devote a deal of attention to it. The same thing, of course, applies to people.

UNWANTED GUESTS

Flies

Flies have unattractive personal habits, though, doubtless, Mother Nature has some use for them.

Where they fall down basically is their total inability to recognise the difference between home cooking and cow-pat.

132

They come striding in, fresh from a tiny glut on some unedifying object best left to the darker reaches of your imagination, and pace nervously about on your meat or bread, leaving dysentery-footprints and even, such to which are the depths they'll stoop to, evacuating their little bowels.

The bluebottle, during the summer months, is almost permanently on the look-out for some good dead meat in which to lay eggs. They even drop them through the wider-mesh meat covers. They make rotten mothers too. The answer, therefore, is to deny them access to their breeding-grounds. So keep your refuse covered up. (The black plastic bags are again a happy solution.) And put your food away in the fridge, store cupboard or, again, plastic bags.

Otherwise it's find a good fly spray – spray and run, before it gets you. I'm not a fly-paper man myself – the results are so revolting and I've shied away from those other hanging fly exterminators at which the finger was recently pointed as they exuded a close cousin of a nerve gas, the CS Gas they've used in Ulster. Perhaps they've improved them since, anyway, enquire and if after a spray, also ask if it will do the job required, as many flies nowadays positively thrive on D.D.T.

Mice

If you have a cat, of course, you're laughing. Not that all cat-owners laugh, but then the more I travel life's byways, the more convinced I am that, blessed with a cat or no, we are all in all a miserable bunch at best.

Mice can make a horrible mess of a kitchen, and they smell.

And they seem to be getting increasingly sophisticated. Many the mousetrap you find next morning cheeseless and unsprung. One trick apparently is, owing to their keen nose they can suss out a human trap at a good yard or three, so put on gloves to set it. It also eases the pain as it closes on a slow-moving finger.

Otherwise, as with the objectionable blow-fly, don't give them anything to survive on. If you don't leave them breadcrumbs about and tasty morsels, they will off to pastures new. Also, fill up any holes with Polyfilla. There are a variety of poisoned powders you can put down, but the average mouse seems to know all about them, and steers a wide berth.

Rats

They are very nasty and carry Bubonic Plague and God knows what. Admit to yourself that some things are bigger than both of us. Send for the rat-catcher. I heard tell of a lady in Australia who happily fed them, thinking they were bandicoots. Do not be deceived, there are no bandicoots in Great Britain, despite appearances.

Woodworm

They relish your woodwork. They can't get enough of it. Plywood is a particular favourite, but they're not particularly fussy. If you put your ear to those tell tale holes, it looks as though the sideboard has been pelted with tiny shot, you can, on a clear day, hear the little bleeders laughing.

Ask the man in the hardware store for a special potion to pour into or brush over the holes, but it's very hard to know how bad the damage is. I remember tripping merrily down a Notting Hill staircase, and three steps crumbled suddenly beneath me, leaving me prostrate on the stair carpet, swinging over the gap as if in a hammock. I had them repaired, sold the house, and mentioned idly to the new owner that there did appear to be some evidence of woodworm. He was an Etonian and not given to panic. 'Have no fear', he smiled, 'Rentokil will put them to the sword'.

While this conjures up delightful pictures of D'Artagnan-like Wormeteers wiping out all opposition with rapier thrusts through the holes in the woodwork, it also makes sense. There are other firms who specialise in woodworm cures. Find out who they are by looking in your local Yellow Pages or classified directory. Compare prices and then leave it to them. Or, again, move.

134

Fleas

As it put so charmingly in an 1898 *Book of Home Management,*
'No one can hope to be free from a few fleas now and then, but to
submit to bugs is a dreadful idleness'. How very true. The flea
is after your blood. Invariably you can blame the cat or the dog,
apart from anything else, that's what they're there for, to be
blamed for most of life's misfortunes and kicked in those
moments you need cheering. So keep the cat and the dog flea-
less. Dogs can be dusted with D.D.T. powder. Ask your pet shop.
Cats can't, because they'll lick it off, but again the shop will
provide some powder, or even an impregnated collar you can
put round their necks. If you've got fleas – and some people alas
are more susceptible than others – thoroughly purge and fumi-
gate yourself, and everything and everywhere round you. They
tend to breed in the floorboards. The dust is a comfort to them.

Wasps

It may be something personal, my best enemies have never
told me, but I have never been attractive to wasps. However,
I spoke to one who is, and here is his tale:
'There was this wasps' nest just by the back door lavatory.
It was growing day by day. The wasps thundering in and out
like Boeings. They were quite friendly wasps – they didn't sting
anybody – and the cat didn't even mind them. Well, when it had
reached the size of a football – the wasps' nest, not the cat – it
was clear that something had to be done about it. I stopped
the children poking it with sticks, told the local pest control
people – they said it wasn't their province, so I shopped around
and found a man in the Yellow Pages. He was large and bearded,
about 25, came in with a sinister-looking black case. He was
muttering under his breath, as he looked at the nest. 'Bastards',
he said, 'bastards'.
'Shut the doors and windows', he said, 'and shut me outside.
I'm going to make some smoke.' He did. I've never seen so
much smoke in my life. The wasps zoomed in and out of this
white fog. It seeped under the doors and through the cracks in
the windows. Three flights up, my eldest son reeled down from
his bedroom coughing horribly. He was the beards' first victim.
There was a great flailing of arms outside the kitchen door –
visible through the glass. He was probing at the nest with a
garden fork, and jumping up and down on it – it was squashy
with white wasp-grubs. 'Bastards', he shouted, 'I hate them!
I hate them!' Eventually the smoke cleared. The wasps were

still flying about looking for their nests. He wiped his brow.

'Do you often get stung?' I asked.

'Hundreds of times', he said, 'I hate the bastards. They won't bother you again. At sunset they lose their way anyway. And they won't be able to find the site of the nest.'

'What happens to them?' I asked, feeling some remorse.

'I dunno', he said, 'and I don't bloody well care. Bastards.'

I gave him £5.'

Ants

Usually come from outside and it's not particularly difficult to find out how. Having found their entrance, fill it regularly with borax or D.D.T. powder, until no longer plagued. If you can find their nests, even better. To avoid ant trails across garden and kitchen always wipe up spilt food which might attract their afternoon ramble. Be ruthless with kettles of boiling water. If you believe in reincarnation, steel yourself and keep pouring, they're unlikely to be anyone you knew.

THE PERFUMED GARDENER

Self-sufficiency is the cry these days. Grow your own. Utilise your garden or your window-box. I used to fantasise about turning a third floor flat I had in Chelsea into a small-holding. Corn idly wafting in the living room, the lowing herd winding slowly o'er the Wilton, complaints from the neighbours at harvest time and forgetting to put the cat out prior to crop-spraying. 'We plough the fields and scatter quickly to avoid the effects of the crop-spraying.'

Food isn't what it was. Soya mince, man-made fish, hand-knitted marrows and artificial tomatoes made from old newspaper. Quite tasteless but jam-packed with protein and useless information. Meat analogues and substitutes culled from fungus.

You're quite right. Grow your own. There's no flavour on earth like a lettuce you have personally raised from infancy. And whereas in a perfect world you would have greenhouses and rolling acres and Old Adam erupting all over the gladioli, lettuce *can* be grown in a window-box. The information on How To is easily obtained but here is an inspirational list. Mulch on, Macduff.

Tomatoes You can easily buy tomato plants – already potted,

and there is one variety – Tiny Tim – specially designed for the window-box.

Runner beans Excellent for a small garden and decorative, blessed as they are with bright orange or scarlet flowers.

Globe artichokes are decorative as well. You can grow them in a flower bed.

Sweetcorn and potatoes are worth having a crack at. *Spinach* too.

Herbs are well worth growing, and will grow almost any-where. Chives, chervil, mint, etc. Try one of these pretty terra-cotta pots with holes for them to emerge around the sides.

Go to one of those gardening centres and make a friend. Heaven knows, my fingers aren't green, nicotine-stained per-haps, but not green. However, one year, I bought some pumpkin seeds, and simply followed the instructions on the packet. Before you could say Percy Thrower, a vast, Triffid-like plant was attacking the neighbour's upper windows, and the ground was thick with pumpkins like medicine balls. It's a pity I hate pumpkin.

We bought a donkey, Miss World, to keep the nettles down. Alas, it transpired to be allergic to nettles and used to stand bellowing among them, till someone hacked out a path to safety. Still that beast stands alone as an admirer of the singing voice. I used to lull her to sleep nightly with selections from *Oklahoma* and the like.

If you haven't a garden or a window-sill, try the town hall and see about an allotment. The amateur gardener is never short of information. Apart from the television, and there's scarcely a

I THINK I'VE BORED IT TO DEATH

programme that isn't growing something, there is a plethora of books and magazines. Or indeed ask around the allotments what they're doing next week or why your asparagus hasn't surfaced yet, while theirs is burgeoning.

Indoor plants

A few good trees around the house or flat are very pleasant, and provide for the lone man, something to talk to. 'Good morning, Tree, and how's the world with you?' and you'll swear a leaf or two wave back.

Keep their earth moist.

Make certain they're adequately tubbed or the carpet and floorboards start to disintegrate and the flat below is in imminent danger of falling timber.

Pull off dead leaves, and have a ruthless prune on occasion. They enjoy that if you keep the chat up.

Quite often the reason the leaves are browning at the end is that they're touching the wall. I don't fully understand that, but move them away.

Your local tree surgeon, or the shop where you got them, will tell you if they need sun or not, or the best part of the room to keep them.

BOBBING AND WEAVING

The important thing to remember if you suddenly find yourself in the ring with Muhammed Ali, and he'll fight anybody, is not to waste your time trying to hit him, but to make bloody certain he doesn't hit you. This is also true of living today. Hence the bobbing and weaving. Don't just stand there – duck!

What causes those moments of anguish when you sit head in hands, moaning alone, and listening to your ulcers growing. Is it money? The thought of your first coronary? Your racking cough? The fact that you keep falling over in public houses? The constant jangle of the phone? The rent man? The fat girl with acne at the cash desk in the supermarket? The invitation to fight Muhammed Ali in Stockport?

Can this book help?

No. Well, I suppose you could throw it at him, but it will only delay the end briefly.

Nor is it bullet-proof, as advertised. That was a misprint. But come along, sir, pull yourself together. I forgot, pull yourself alone. Never say 'die'. It isn't meant to be easy. There, you see, you're bobbing a little already.

DIETING

There comes a moment in almost every life when one contemplates one's navel and suddenly realises it's a good foot further away from you than it used to be. Mine almost invariably was. I've inclined to stoutness all my days and like all those who get used to being fat from an early age, could be hailed as 'Porky' or 'Fatso' without taking offence, developed sturdy legs (hollow too, such are the wonders of Nature), put it all down to some mild glandular malfunction, learned to live without being able to touch my toes, and studiously ignored the fact that I was pouring beer, bread and potatoes down me like some giant waste-disposal unit.

It shows how much I've taken my belly for granted over the years, it was only recently when writing a piece for a slimming magazine (only one calorie per page, but quite nutritious) that I discovered I'd always spelt it wrongly. Not that I'd ever written to it. But I repaired to the dictionary and looked up 'stomoch' and was sharply put in my place by that stunted but learned fowl, the concise Penguin, which squawked quite clearly:

'Stomach (stumak) a membranous sac in the abdomen in which food is digested – PDB: belly, abdomen, (fig.) appetite, liking, wish; (at) temperament; haughtiness or stomach v/t digest; eat with relish; put up with; tolerate.'

All human life is there. I find it hard to think of the dear, old thing as a 'membranous sac' but by Heavens and upon my oath, I've certainly eaten with relish and without, and, old Stomach, you have certainly put up with, not to mention tolerated, some vile abuses.

It's that stomach you've got to watch. That's where the trouble lies, and if you're thinking to yourself how best can I rid myself of it, there is only, alas, the hard way. Hard, but in the end rewarding.

What you have to do basically is the *shrink* that stomach so that you cannot physically consume the amount you used to. This you do by a mixture of diet and exercise. The one stands no chance without the other.

If you have a serious weight problem it is wise to consult your doctor.

Now there are books that suggest that by eating grapefruit before meals, or eating only fat, or drinking six bottles of whisky a day, or living entirely on quail's eggs and marmalade, or Pernod and radishes, you will become sylph-like in a nonce, and there may be an element of truth in all of them but in their own way they're conning you into the mistaken belief that the road to thinness is paved with fairy cakes and pints of Old and Nasty. Not so, I repeat, you are going to have to make sacrifices, dearie, and by that I do not mean appeasing the deity by slitting the throat of the nearest goat.

142

143

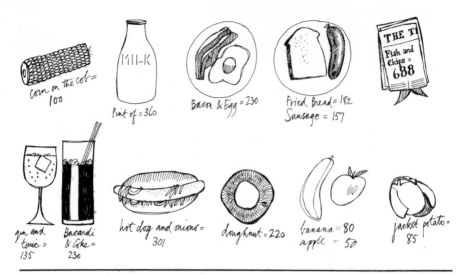

Corn on the cob = 100

Pint of = 360

Bacon & Egg = 230

Fried Bread = 182
Sausage = 157

THE T̶
Fish and
Chips =
688

gin and
tonic =
135

Bacardi
& Coke =
230

hot dog and onions =
301

doughnut = 220

banana = 80
apple = 50

jacket potato =
85

Sacrificial rules for slimming

1 At that recent moment of truth when you faced the belly, did you not at once breathe in deeply, perhaps for the first time in years, and mustering what muscle remains in that area draw in the great bag of lard and try and lift it chestwards. For one fleeting second, Charles Atlas lived again, didn't he? You thought of glossing up the body with baby oil and waiting for some poor fool to come and kick sand in your face.

Take heart, and incidentally, breathe out immediately before you collapse in an undignified heap.

What you did just then was ideal and quite the simplest way of re-awakening old stomach muscles. And you can do it where'ere you walk. Or sit.

2 So that these aching muscles don't finally snap under the strain, the second thing to do is stop distending the inner man with such vast quantities of food and drink. It's Archimedes' Principle in reverse – drink a bath full of water and you swell up proportionately – Eureka!

Cut down, starting from now. (To tell the truth, I addressed that more to myself for as I write I sneaked a glance downwards, and realised that sad truth that whatever comes down, must go up. To think one happy summer this previously 16-stone weakling whittled to 12 stones. Those dear, dead days before I relaxed and started slowly climbing once more. At least now I know it can be done, but, oh, it's the private agony. I too shall steel myself.)

spaghetti bolognese = 370

heart = 180 (4 oz.)

= 160

black coffee without sugar = 0

pint of bitter = 180

boiled egg = 80

2 fish fingers = 120

1 carton Fruit Yoghurt = 150

apple pie & custard = 360

525 Kentucky Fried Chicken & Chips

baked beans on toast = 180

3 *How?*

a Go out and buy one of those simple guide books that tell you how many calories there are in a cauliflower cheese, etc. At the end of this book are listed some of the basic daily delights.

b Ration yourself to 1,400/1,800 calories a day. Cauliflower cheese is .round the 200 calorie mark. So eat six cauliflower cheeses. There's probably a book that recommends just that.

This ration allows you 9,000-10,000 calories a week or 35 doughnuts. You are paling visibly. Scrawl old adage on wall – 'It's not meant to be easy'. (This applies not only to dieting, but life as well, and I find it a comforting thought.)

c Surround yourself with as many low-calorie commodities as possible, the special low-calorie margarines and breads, etc. This will give a sense of dedication and mild saintliness. I can also recommend some of those biscuits you can buy in chemists and down with a glass of milk (360 calories to the pint). They stave off ravenousness, and press further home the point that suffering is involved.

d Keep your regimen simple. You can get books of day-to-day recipes advocating change, but I reckon you merely become confused, and, anyway you suddenly find you're breakfasting on stewed fruit and yoghurt, which may be *your* bag, but fell out of mine.

e Keep taking the vitamins. Look what they've done for Cary Grant. He may rattle, but he's still pretty. I've come to believe in them wholeheartedly. To think, a year ago from now I qualify for Phyllosan. I'll be a fortified over-forty. I wonder if there's some sort of presentation. I use Pharmaton, which contains everything and ginseng, the Chinese panacea. The easiest way to take them I find is to ask your chemist, always a useful man to know, for a multi-purpose, multi-vitamin-packed pill and have it with your first cup of coffee. For the first month of a diet, you feel weak at the knees anyway and the vitamins do help.

f Set a target. Funnily enough I aimed for a measurement, rather than the loss of the weight of a leg or half your baggage allowance on Concorde or whatever. I'd always fancied a pair of red jeans, but if you ever look at a rack of trousers in the Men's Department, it is a dismal prospect for the larger person with the more mature body. Down around the 28-inch waists, the early 30s, all the colours of the rainbow jostle; reds, yellows, indigos.

Bright checks that Nature never hazarded. But as you scan the spectrum towards the broader girths, the sun goes in. Autumnal tints, brown and seedy beige hang somberly. Around the 40s winter. Grey and black. They hang like baggy storm-clouds. The times I'd seen in King's Road windows, gaudy trousers that suited my mood, but never got further up than my shapely knees. 'You're very *big*, dear', frail, young assistants would say, bedecked in trousers that would cause a cat to turn and flee.

g The other target. Think of a weight you would ideally like to be or, safer, consult a height and weight chart, and subtract your ideal weight from what you are. If it comes to say – 40 pounds – think instantly of 40 pounds of lard or butter and how much space that occupies in your local delicatessen. And indeed the weight if you decided to carry it home. It's about the weight of a three-and-a-half-year-old kid and how far can you transport one of those. You will almost instantly come out in sympathy with your legs, and wonder idly how they've succeeded in walking this far.

Keep weighing yourself. At times this can cause you to look at your vows afresh, when your stomach's growling with hunger, Your hands shake with ague, and not an ounce seems to have budged since last Wednesday, but, ah, the bliss when suddenly you hit a target. Three pounds in a week! The spirits soar. Celebrate with a hard-boiled egg (80 calories) and a glass of dry, white wine (70 calories).

Health

When you do lose weight, you feel so well. Hours are added to your day. No more starting awake in your Parker-Knoll, and realising with a shock it's time you got up and went to bed. No more feelings of acute depression when you see those advertise-

146

OO!
AREN'T
WE
ENORMOUS?

147

ments of happy men dancing on beaches with a large 'X' on their chests, as though they had been found to suffer from Dutch elm disease.

Think thin and go for your life

It's a voyage of discovery. You discover things like cheek-bones and ribs. I've jotted down herewith a few spare-wheel puncturing notions; but, in the main it's up to you and a calorie-counter, and your doctor too if you think your constitution won't stand it. Don't point the finger at me, if you keep falling over.

1 If you've gone too far the previous day, take it out on the next one. A day of total starvation has amazing results, even if it hurts.

2 If you get extremely hungry, the willpower begins to ebb so do not hesitate to chomp occasionally on an apple (50 calories), a tomato (10 calories) with a slice or two of crispbread (around the 35 calorie mark). I found cold roast beef saved the life quite often, with a glass of wine.

3 Alcohol is fattening, but then so are soft drinks: Ginger ale 100 calories – Half pint of bitter 90 calories – Glass of orange squash 120 calories – Small scotch 75 calories. So count away at those calories, knowing your daily allowance, and the odd buzz is acceptable.

4 For example, if you count carefully you can have the odd spud, the occasional pudding, and why not?

5 In the main, however, forswear the starch and sugar and let your conscience be your guide.

6 I must admit I have a very thin conscience. If you are thin, why are you reading this at all? People like you make me sick.

HOW TO GIVE UP SMOKING

1 Stop putting cigarettes in your mouth and lighting them.

2 Stride about, boring your friends loudly with cries of 'I have astonishing willpower, for – years I have smoked – packets a day, but from this moment forth never again shall a gasper darken my moustaches'. This makes it a matter of pride.

3 Always carry an unopened packet of cigarettes on you. Then you can look at them and say 'No, I have given you up'. If you haven't any cigarettes on you, then the only reason you're not smoking is that you haven't any cigarettes on you.

4 Chew pencils rather than boiled sweets. There is no substi-tute, alas, for smoking. You must think of other activities for this

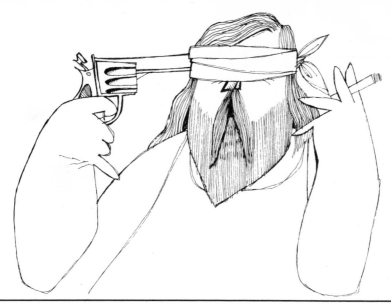

new spare hand. For those moments when you used to push your work aside, sigh deeply, lie back and light up, than the which no moment is holier, think of some alternative relaxer like firing an air pistol at your least favourite books.

5 Never think of yourself as a non-smoker. Non-smokers have never known the joys of it. You are a smoker who isn't smoking.

6 Try one of these mouth-washes now on the market that make cigarettes taste like spider-droppings. This seems most effective unless you have a penchant for spider-droppings.

7 The withdrawal symptoms last six months at least, sometimes a year. But in 20 years you should be over the hump, and your chances of lung cancer lessened.

8 Fall under bus and serve you right.

9 Have a cigarette.

HOW TO GIVE UP DRINK

I did it once for a year and a half. This proves conclusively that anything is possible.

Alcoholics Anonymous have a sensible attitude. They don't suggest you give up for the rest of your life. That in itself is a sobering thought guaranteed to drive you to drink. Simply say to yourself 'I shall not have a drink today'. Tomorrow can wait.

1 Go to the sort of party of which you used to be the life and soul. Drink orange juice. Watch old friends become strangers. See strong men stagger and fall. Hear the same story several times from famed raconteur. Leave early before the vomiting starts. After all, you can only drink so much orange juice.

NOT ANOTHER
DROP SHALL
PASS MY
LIP

ELECTRIC
SOUP

2 Find reasonable substitutes, *(a)* China tea – it has the bitterness you'll miss. *(b)* St Clements. Very popular in the BBC Club with drying-out producers and executives. A bottle of orange and a bottle of bitter lemon stiff with ice and any old peel they can muster. *(c)* Lemonade and lime juice. *(d)* Ginger ale packed with ice and slices of cucumber. *(e)* Any French mineral water you like.

3 Find out how much easier it is to do things (addressing the Oxford Union, attacking builders, suggesting that the lady in your life may be slightly wrong) without the three large ones under your belt you always thought necessary.

4 Ignore shouts of 'You miserable bugger'.

5 After a year and a half start again – it's more expensive now but you don't need so much.

TRICKS WITH A TELEPHONE

To keep calls short, pre-script them. If you have three questions, say, to ask of someone, write them down on that highly necessary pad that should ever lurk by the phone, with ballpoint or whatever.

If the other party will not get off the line and you're paying, start dialling, this produces eerie silence and suggests breakdown in communications. In extreme cases, dial and hold finger down, slowly replacing receiver before releasing dial. Other party will ring back at once. Do not answer. Conversely if you

have more to say, remember that they are now paying, so answer. If you didn't answer and were caught napping by the other party later in the day, say that you had gone out to telephone the engineers.

If you seek total peace and require no interruptions, dial '2' and jam a pencil into hole '1' which you will see has just passed the tiny tit at the bottom. This kills the blower temporarily.

If you are anticipating a call you could well do without, adopt tactic above or answer, pretending to be machine. Give your number (think Dalek during this impersonation), and simply say 'This is a recorded message. Joe Stoat is (a) not in at the moment, (b) away for some months, (c) dead. Please speak now. Bleep. Bleep. Bleep.' This way you discover painlessly who it is and possibly what they want.

The anticipated call you could well do without has got through. You weren't thinking. A trick used in the early days of *Private Eye* when faced with telephonic heckling or abuse leaps to mind. An accomplice is best, but it is possible to work it single-handed. One of us, at a given signal, would pick up an extension and in a muffled voice say, 'Sorry to interrupt. Engineers here. We've had a spot of bother with this line. While we're checking it out perhaps you would care to listen to some light music played by the Post Office Ensemble'. And, at that, clap a transistor radio to the telephone. Collapse of Other Party.

If you can afford it, actually get an answering machine or answering service.

I frequently pretend to be a Peruvian au pair girl. Do not be deceived.

SHOPPING

Some basic rules

1 Keep a list going on a slate or board in the kitchen which you can translate on to paper when embarking on expedition to shops.

2 Never shop in a supermarket when you're hungry.

3 Never use a trolley. A wire basket is a fine insurance against casual acquisition of excessive and unnecessary tins and bottles.

4 Stick to your list. Do you really need that 3lb tin of American stuffed cacti at £2.80?

5 Try and pay in cash. It keeps you in order. It *is* possible to live entirely on 'tick' – our Great Nation does – but it don't half catch up with you. Look at our Great Nation.

6 When shopping for clothes – take a lady with you. She will feel deeply flattered, and have someone else to blame if Papal Puce isn't your colour. Apart from which, she's more likely to shorten the trousers for you if she thinks she helped you choose them. Don't forget the Oxfam winter collection. Particularly good for that oh-so-nostalgic look. Ex-Catering Corps greatcoat, gas-mask and puttees.

7 Only buy easy-to-handle fabrics, colour-fast and the like that won't disintegrate or fade away like old soldiers at the launderette.

8 Shop around. It's bound to be cheaper somewhere else.

9 Keep your receipts in case you have to go to war.

10 Remember you are not alone. Shoppers are quite well-protected. You don't actually get someone from the town hall riding shotgun on your wire basket, but there's someone there who'll join in the fracas if you feel you've been done by a shop.

For your protection

The Trading Standards Department They handle Weights and Measures, the Trade Descriptions Act, they're the people to contact via the town hall if you think you've been conned by a shop or the advertising.

Consumer Advice Centres These work hand in hand with the Trading Standards Department, and will advise you on any queries that come up before or after buying. They'll also check out any hideous form you're called upon to sign.

The Old Public Health Inspector now re-christened the *Environmental Health Officer,* which doubtless makes him feel better, is equipped to descend with heavy steel-tipped boots on any shop, restaurant, massage parlour, etc. that strikes you as so unhygienic and foul that you have to tell somebody. Tell him. If you're uncertain in your wrath as to which to turn to, simply ask your local *Citizens' Advice Bureau,* they will straighten you out. Free of charge, too.

A case in point. You buy a pair of trousers and after only a mile the crutch explodes. Now the shop may do two things: *(a)* offer to heal the exploded portion or *(b)* give you a credit note. Either of these offers may blow your chances of getting your money back later. You are perfectly within your rights to demand a new, bomb-proof trouser or your money back.

CAVE-DWELLING MADE EASY

It hasn't actually come to it yet, but I have a feeling that cave dwelling is the trend to watch for the future. I've been eyeing a few in north Wales to retire to when the bottom falls out of Britain as we know it.

In the meantime, you are doubtless ensconced in a room, a flat or a house, depending on your preference or wherewithal or wherewithout.

Gone are the days when an Englishman's freehold was his castle. Indeed, the feudal baron would now be infinitely more securely protected in a rented, furnished lodging than housed in his Norman keep, from which he would be quite likely to be evicted on the payment of miserly compensation, by the local

153

TECHNICALLY SPEAKING, THE RESIDENT LANDLORD IS A STEGOSAURUS

council to make way for a road. Boiling oil is a thing of the past. Indeed as more and more residential areas give way to the motor car and juggernaut, the simple solution seems to be to buy a caravan and take to the motorways of Britain, driving round and round in circles and sleeping in shifts. It has advantages – the Inland Revenue will never catch up with you unless you pause too long in a lay-by.

If you're renting

The 1974 Rent Act was born in August 1974, which makes it a Leo, like me, masterful, ambitious, obscure and invariably long-haired and bearded. Hear what comfortable words Shelter sayeth: 'The Act was passed by Parliament in a rush before the summer vacation and as a result the wording is obscure in places'. I warned you, a typical Leo. 'The meaning of some sections will only become clear after test cases in the courts.'

If you are now paling visibly remember you have on your side the Rent Officer and the Rent Tribunal, the good old Citizens' Advice Bureau without whom . . ., the town hall who should put you in touch with the right people for your case and, of course, Shelter.

Ask yourself these questions

(a) Am I living under a resident landlord? *(b)* Did I move in before or after August 1974? *(c)* Am I a fixed-term contract or a periodic tenancy? *(d)* Can he hurl me on to the streets? *(e)* Am I paying too much rent?

154

Answers

a A resident landlord is someone who has decided to rent off bits of the old home, and as his title suggests, still lives there, probably in the fanciest bit. In other words, the building wasn't designed as a block of flats, and you know him at least by sight (which isn't as silly as it looks, because you can find yourself paying rent to some weird bodies and companies). You're *not* protected, furnished or not. *But* if you think for one instant, ho, ho, he's hardly ever here or I fancy this is not his sole abode – check up, and you could become a protected tenant.

b Did you move in before or after 14 August 1974? Because you *are* protected if you had an unfurnished tenancy that began prior to the Act, as long as you don't share a kitchen or bathroom with him or her. (That doesn't include social occasions, naturally.) If it was a *furnished* tenancy before August 1974 you're *not* protected. There is a case, however, for checking up whether it was sufficiently furnished to be called furnished. Some fast ones have been pulled.

c A fixed term contract. You have to go when it's up, but you're protected during it.

A periodic tenancy. This can only end with a Notice to Quit delivered by either side. Four weeks is the minimum notice you can get, and it lengthens with the length of this let. You're not protected *but* –

d No, he can't hurl you on to the streets, whether in fact you're fixed term or periodic, without a court order, and he's up before the beak if he leans on you, viz. cuts off your electricity or sends large men with violin cases round to put his point of view. If you're a protected tenant, and you have paid your rent, and haven't attacked the neighbours or smashed the place to a pulp you should be as safe as a house. (Not a happy simile.)

e The rent? If you're *protected*, and furnished or unfurnished really doesn't matter any more, ask the Rent Officer to register a fair rent, if you think you're paying too much. Once that rent is registered, it can only be upped within the next three years, if either the landlord improves the place or if the rates go up.

If you're not protected you can go to the Rent Tribunal and they will set some reasonable sum for perhaps six months at a time. You may, on the other hand, be the sudden recipient of a Notice to Quit, so ask someone first.

The obscurity may alarm you, but there are excellent pamphlets to be obtained from the town hall, public library or

Shelter and always somebody who'll help, including a friendly solicitor. (Read *Regulated Tenancies – Your Rents, Rights, and Responsibilities* – it reveals all in quite readable terms.)

If you're a landlord

And I mean a resident landlord, which is quite a sensible thing to be in these dark days, if the house is too large, if you're lonely or if you need the money.

I have heard of a gentleman who gives away rooms but only to those who will give freely of their services to him. For example, an accountant, a solicitor, a couple of girls for hoovering, washing up and baby-sitting. I believe he is still looking for a plumber, an electrician and a vast, blonde nymphomaniac as advertised on page three of *The Sun.* This seems very wise. Well, up to the electrician, anyway.

Suffice to say, there you are in your own place and you want to rent off bits without losing control.

a Base the tenancy agreements on a weekly or monthly basis, these carry on blithely until one of you comes up with a Notice to Quit.

b Don't get into fixed terms in case your life-style changes.

c For weekly tenants you'll need a rent book.

d Unless tenants have been there since before 1961, you are bound to keep the hot and cold water, gas, electricity, lavatories and heating going.

e If you're tarting things up for your tenants, try the council for an Improvement Grant. You never know.

f You can get Notices to Quit from the council or a legal

156

stationer. It has to be worded in the correct parliamentary manner and you'll never capture the balls-aching style on your tod.

g Having served it, and the appropriate time having elapsed, and the tenant not having budged (this reads as is translated from the Latin) – what to do? It is, alas, time to come the heavy landlord, twirl the black moustachios, adjust the silk topper, and via the county court summon the sheriffs and bailliffs.

Owner/occupier

That's me, and I've owned and occupied a few. My personal philosophy is keep moving, because you present a harder target, and bobbing and weaving is the only exercise I get. Apart from coughing, (see Getting the Lungs Working).

Nowadays, freeholds, unless you have one already and are grimly hanging on or landlording it about, are without the means of most. Leaseholds are favourite, if you can raise the necessary, think of them as quite a modest rent but paid in advance. For example, £10,000 for 10 years in a four-bedroomed flat in central London may sound a lot, but look at rentals in *The Times* classified ads to see what they come out at monthly. A leasehold allows you to live slightly above your station. I could never have got near the freehold price of my flat. If it's over a 50-years lease, you stand a good chance of a mortgage. Under that, they seem unwilling, which is foolish. It's not an unsound investment. Even a 10-year lease.

Get a solicitor, though, to comb that lease. If you're hazarding a freehold get a solicitor and a surveyor. It costs a bit, but it's worth it, and you have some redress if the bath falls through the billiard room ceiling. Fortunately I'd sold that place a week before it happened. (When I say billiard room, actually, it was a bedroom with a billiard table in it – you can sleep anywhere, but billiards needs space. Even so, you had to go out on the balcony for some shots.)

Moving

I have become something of an expert. Good, frequent shifting of domicile does two things for you (a) it keeps you going financially if you're relatively sensible – and (b) it keeps your personal rubbish down. It's the one occasion when you can get rid of all your useless paraphernalia. Last move I whittled down the books by 50 per cent. The one before, furniture, rugs, etc. Auctioneers will take almost anything these days.

157

158

Rules of moving

1 Pay up and have cut off newspapers, milk, telephone, electricity, gas, water.

2 Get the 'pros' in when it comes to removals. Get some estimates, and let your instinct do the rest. You can do it yourself with a Rent-A-Van or whatever, but hernias loom.

3 Ask them to drop some packing cases off a few days prior. There's a lot you can shove in yourself. Bedding, clothing, books, kitchen stuff – I'd leave the china to them, they have the knack.

4 If you have anything fragile, lovely or irreplaceable – take it yourself in the car and save on heartache.

5 Anything too large, but fragile, lovely and irreplaceable is best sold – then you will never suffer again. Otherwise, point it out to the oldest and wisest of the removal men, find out if their insurance covers it in toto, and leave it to their tender ministrations. I told you you should have sold it.

6 Label all the boxes and then you can stand at the door of the new abode and direct operations.

7 Find out the parking problems at the other end, so the fuzz can be warned. Traffic wardens will quiver and say it's more than their job's worth to let you park there longer than one minute.

8 Get the local Post Office to send your mail on for a year. By that time all those who communicate with you should have gathered where you are.

9 Go to a hotel or friends for two days thus avoiding most of the chaos.

INSURANCE FOR THE SINGLE MAN

The last thing you want as a single person of male inclination is one of those insurance schemes that pays out a wonderful sum of money at about the moment they're patting the sods securely down on top of you so you'll keep longer – unless you have relations, of course, that you don't mind enjoying yourself more than you did.

The best thing for you is either a good *Pension Scheme* so that you can grow old disgracefully, and even if your firm provides one you can still underwrite it and show more of a profit. Better still, an *Endowment Policy*.

This costs more up front, but if the policy is for, say, 10 years, then after that decade you get the maturity figure *and* bonuses

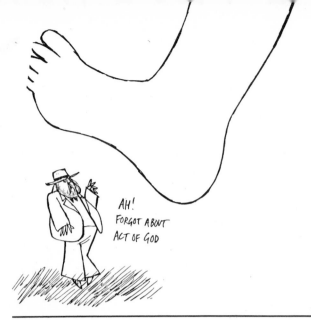

AH!
FORGOT ABOUT
ACT OF GOD

---

and, good for the single man who is universally clobbered by the Chancellor, tax relief.

(Let me ask you something – how on earth does anybody have any money left about to indulge in all this? Wise men will talk of savings and the advantages of building societies . . . What savings? I cry. They are silent now.) *Moral:* If like most of us you aren't shooting the rapids up Shit Creek, but are in stiller waters with a paddle handy, find an honest broker.

Insuring Property
Insure the flat or house just in case it happens to you.

Articles of value such as cameras or tape recorders, wirelesses or electric harmonicas can be insured separately and it gives them automatic cover for the first 30 days in Europe or round the Mediterranean. (If you go anywhere else with your electric harmonica, you'll have to get another quote and pay a bit extra.)

Travel Insurance
Again worth it in these troubled times. God knows what hideous privations you may not suffer in foreign parts. Cover yourself against looting, pillage and rape.

A word in your ear
To tell the truth I have now whittled down my possessions to so few and valueless that the last time I was burgled, the thieves very kindly left some objects for me. I still had to pay for a new door though, so it seems wise to insure.

160

A funny thing

The honest broker I spoke to about insurance and the single man told me how once a person approached him and said he wanted to insure his coupé.

'What year is it?' enquired the H.B.

The man was clearly non-plussed.

'It's brand new', he said.

'What c.c. is it then?' pursued the H.B.

'C.c.?' queried the person, his jaw tangled in disbelief. 'I want to insure my toupé.'

The happy outcome of this little tale is that the person lost it some months later running for a bus. There is a moral there for all of us. Particularly you, baldie!

A TRAVELLING MAN

Whichever way you are travelling – on business trips or holidays, by plane, boat or the simple motor car, the single man travelling alone is invariably blessed with advantages over his fellow man, lumbered with spouse, kids, cats and no space to spare for those unseen delights which may well fall into his path/arms at any point en route.

GETTING PACKED

Getting it all together first before you can even depart can be a problem though, and the most boring aspect of travel is the *packing,* so let us first get that behind us, where it belongs.

What you need to consider is where you're going, what for, and how long? If it's a 14-day holiday in the Algarve in August – don't pack your tweeds, don't pack 14 pairs of socks you'll rarely wear them and don't forget your dark glasses. On the other hand, if it's a 14-day seminar in Bradford organised by your firm, the Kamikaze Laxative Co., and it's October, pack the tweeds, 14 pairs of socks, and forget the dark glasses.

Rules for packing

One extremely *heavy suitcase* is better than three light ones. You're less likely to lose it, and there's a fat chance of anyone stealing it.

Imagine yourself dressing and lay out the items in reverse. Or imagine you're undressing and lay them out in that order. That seems easier. Now pack in that order.

162

Shoes will therefore go in first. This is excellent as the heaviest things should go in at once. If on business abroad make certain you have shoes that will last. They're hell to buy elsewhere.

So as not to waste space, put socks in the shoes or other small objects like electric razor, alarm clock, etc.

Socks I work on the one pair a day principle. It's safest. If in a hotel and they're evil-smelling and you can't be bothered to wash them, pop them in one of those bags so tastefully provided for ladies' used euphemisms. If not in a hotel, wrap in newspaper or plastic shopping bag and put back in suitcase.

Shirts, underclothes, etc. If you're on business and staying at a hotel see if the hotel will launder them for you and it can be charged up in the foulness of time. Take all your dirty ones too and get them laundered on the bill.

Don't forget that 'just in case you need a tie' tie, if you don't usually wear one.

Night attire I stick by my affection for the night-shirt but take whatever turns you on. A dressing-gown is a useless thing unless it's a towelling one, or needed for flitting across hotel corridors and other such occasions, when it comes into its own.

Sponge-bag Some sort of waterproof, disaster-proof bag for the toothpaste, toothbrush, etc. Shampoo in a plastic bottle; soap (a good-size bar, as trying to work up a lively lather with those tiny little hotel bars is laughable). You can obtain tubes of travelling detergent. Do not confuse with toothpaste. (I think the reason I grew a beard was that I was constantly cleaning my teeth with shaving cream.) A sponge or flannel. A nail-brush is rarely provided.

Coats, trousers and waistcoats Folded, as illustrated, and to fit your case.

Of course, to save yourself bother, buy one of those suitcases you can hang up your clothes in. They're double-sided, having a pouch outside for shirts, etc. and plenty of room for shoes.

Now pack down the sides impedimenta such as bedroom slippers (a pair of plimsolls will serve and are more versatile), books, hair-brushes, comb, your brand of fags (if wherever you're going seems unlikely to have them).

Destination extras

Now that your basics are coped with, prowl round your room, thinking, 'Where am I going, what for and for how long, what are the likely hazards?'

163

Lay coat down
flat – inside out.
Turn collar up.
Slip hand in and
(a) push both shoulders
   out
and (b) see that
the sleeves are
flat inside

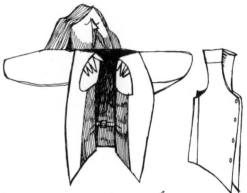

Fold waistcoat in half – thusssish

½ way
mark

↑ waistcoat

(a) Lay trousers out flat
(b) Lay waistcoat to ½-way mark
(c) Lay jacket to ½-way mark
(d) Fold bottom of trousers back over
    jacket and waistcoat
(e) A bit of jacket collar will over-lap –
    fold it over end of trouser.

If it's a holiday abroad extras to remember to pack are towels for the beach, flip-flops, jeans and something to swim in, T-shirts and a sweater probably for the evening. A folded kit-bag in case you come back with more than you went with. Passport (I don't know where you've put it). Driving licence in case you want to hire a car.

Years ago I bought an ex-War Department horse-feeding bag, a large, brown canvas bag that will hold almost anything, including the majority of a horse's head, and hangs from the shoulder, leaving, and here's the main advantage, a free hand. Also, it's brilliant when you get there for beach towels, sun-tan lotion, good book, etc.

Equipped with that and a suitcase I can go anywhere.

If it's business at home, it's probably suitcase and brief-case time and a whole new ball game with a clean shirt and change of underwear.

Travelling to a gastronomic wilderness

To wherever you are travelling a hazard not to be overlooked by our gourmet friends is that you may find yourself in a gastronomic wilderness, the effects of which could completely ruin your sojourn however short – so be prepared.

A travelling gourmet kit

A full *pepper mill.* This should enliven most dishes. A bottle of cheap, fruity *brandy.* Perks up fruits and puddings and warms your cockle. Dry *sherry* to deal with the more tasteless soups.

Worcestershire sauce. Apart from making even the most limp hotel breakfasts start up and take notice – it assists the digestion by neutralising the fats so prevalent in the unavoidable British fried dish.

In case of very unpleasant bread and no alternative, an airtight packet of oatcakes. Or ask for toast.

A *Swiss Army knife.* Those who have seen the Swiss Army drilling with these outside the barracks in Geneva will know of their manifold uses. Good blades for cutting salami and gutting fish. (That wasn't my idea, the knife was suggested by a more hardened traveller, and I agreed with the knife. An excellent thing to have about you, I said. For gutting fish, he said. So I've put it in. Some people.) It opens oysters. (Him again.) Tell the truth fish and oysters apart, I wouldn't be without one. Bottle opener. Can opener. Corkscrew. Scissors. Screwdriver. Saw (*see* Remove your own Appendix).

165

To the Plug

For instant boiling water there's a splendid *electrical water heating gadget* which you simply stick in a tooth-mug of water and it boils up in a trice. Keep a tea-bag or two with you, small packets of instant coffee, milk and sugar stolen from planes or hotels and there you are, sitting in bed, supping away, and blessing me for bringing it to your attention.

You can also sterilise your false teeth, or the inside of the tooth-mug. Use it for your hot lemon drink if ailing. Wash your feet without getting up.

While you are away

Insure any cameras, etc. you're taking with you. Insure house and contents if you don't usually, and only leave out neatly for burglars those things you don't want but would infinitely prefer the insurance money for.

Warn the fuzz you're away. They're over-worked, but it makes them feel wanted, and can do no harm.

Stop milk and papers. Why not stop them anyway? Are they good for you?

Stop paying bills. They can await your return. Being away is a fine excuse for not paying them.

Put any indoor plants in the bath with about half an inch of water.

If it's business, keep a diary every day noting where you were and who you met. Apart from the fact that these are pieces of essential information that evaporate in the night, it makes it easier next time you meet to say 'Ah, Henderson, haven't seen you since February in Torquay.'

167

Air travel narrows the mind

I don't care much for flying. It's not fear, so much as the discomfort. Claustrophobia tends to set in on boarding. I get the feeling that I'm being tinned. I have appalling visions of those air-vents over your head suddenly gushing forth olive oil or ketchup and filling the whole can with nourishing, edible people with the added taste sensation of grilled tweeds and broiled hand-baggage.

Be prepared

If it's a longish trip make sensible preparations. Dress comfortably. Bedroom slippers are a great help. A sweater, because it can turn chilly in aeroplanes. Don't wear a stiff collar or crutch-strangling trousers. Plenty of books and magazines. Start your novel. Keep toothbrush and paste handy, you become very jaded and foul-mouthed.

Try and get the seat by the emergency exit, the one with the leg-room. If it's a trip to somewhere like Australia, do stop off for a night or two on the way. It's the only tried antidote to joining the Jet-Lag-Set.

I remember Malcolm Muggeridge confessing years ago, before he was converted on the A4270, how dearly he would have liked to kill an air hostess. If you feel this way, fly Qantas, they have very few and concentrate on male stewards, who, at the risk of sounding chauvinistic, are infinitely preferable, and made of flesh and bone.

Make sure you have sturdy luggage. These new super-computerised luggage dispensers have a disgusting habit of vomiting up half-digested, mangled luggage.

Go by train instead

If it's Europe, or Scotland or relatively close, why fly at all? There is no more romantic way to travel to Gay Paree than on the night ferry – that is if you are taking your own car. You get on at 10 at Victoria, are shown to your bunk, and apart from slight juddering and clanking as the carriage is lashed down on the boat, you bed down happily enough till *le petit déjeuner* which you enjoy at about 7.30 as you coast into Paris. Where else could you have essayed Position 43 on an upper bunk in a train at sea? No point in being in Paris in an hour. Paris is not an hour away. If it was, they wouldn't be nearly as foreign.

168

Taking the car

I'm a devotee of sleepers, particularly the ones you put your car on. A grand start to a short break. I find it easier to hire abroad though, I find the mental anguish of remembering which side of the road you're on, while driving right-handed rather too much. At least when you're driving on the left-hand side the whole business is mirror-image, and once you've got used to not changing gear with the door handle it's infinitely simpler.

It's extraordinary the number of places you can go, where you *don't* need an International Driver's Licence. Check, however, with the AA or RAC.

Most countries have strange quirks and regulations for motoring. In Portugal, for instance (and the Portuguese rival the Belgians for the Gadarene Trophy for the Most Lethal Drivers in the Universe), you have to hoot when overtaking. Again check first. Also, about giving way to the right.

DO NOT ASK A MAN TO DRINK AND DRIVEL

The inner working of the car are a mystery to me and I like it that way. Both the AA and the RAC are very kind and sympathetic. Not only women can appear helpless. I have never subscribed to the popular view that you can tell a man from the length of his bonnet. However the one general rule I adhere to is that once you've found a garage that understands the vagaries of both your vehicle and yourself – stick to it, and meanwhile:

What to keep in the car
The AA Book of the Car
The AA Book of the Road
Some books for you to read while waiting for the AA or RAC
Some kids' books, a toy or two, and paper
A first aid kit
Plastic bags for rubbish, and also for popping over parking meters in cases of emergency
('Some joker', you can say loftily to a warden, and then add, 'I think traffic wardens should be armed' and watch their faces light up.)
A camera.

What to keep in the boot
Tool kit and jack, obviously

A tow rope, in case, and an emergency windscreen

String or, better, copper wire to tie the exhaust pipe up if it collapses under you.

Insulating tape, to stick the lights back on if you've disagreed with a bollard. Also useful for temporarily repairing the water hose

A plastic container full of water

A can of petrol is no mean notion

A crate of Johnnie Walker Black Label, you never know when you'll need one

A football

A lilo

The fuzz

I approached a friendly sergeant about what to do when stopped by the Constabulary.

'Surrender', he said, 'and be terribly courteous'. Leaping out of the car and shouting, 'What the bloody hell – I am a taxpayer', instantly puts their backs up and the notebook is out in a flash.

Other possible gambits

1 If with a lady, stuff coat up her jumper and cry 'The contractions have started'.

2 'Food poisoning, Officer, I must to a lavatory.' As a general rule, bodily functions are always a good excuse as no one likes to go into them at length.

3 'Thank God, you stopped me, Officer' and report in excited tones the sighting of a highly sinister Unidentified Flying Object.

Words of Wisdom

If, remotely knackered and driving on a motorway, get off it, and have a quick kip. Nothing could be more alarming than waking up at 70 m.p.h. hooked on to the centre rail of a motorway and travelling like some berserk tram.

Do not park on a hill when using the vehicle for amatory activity. I remember once idly walking the dog over Winter Hill in Berkshire, and hearing frenzied cries, seeing some 20 yards below me, an upturned car with two persons, one of each sex, peering out, nude and desperate.

Do not hesitate to keep any dents or wounds your car may incur. My old Rover looks like the neighbourhood tom cat, with one battered ear and the scars of a lifetime. It's amazing how other cars, particularly taxis, avoid you. It is also less likely to be stolen.

170

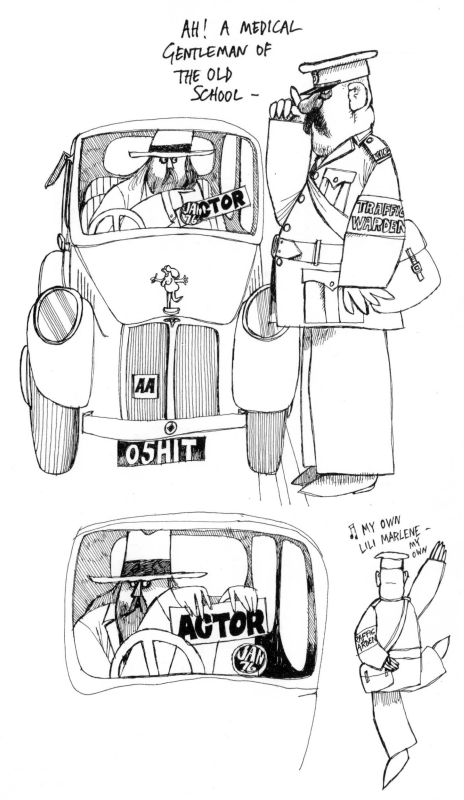

Hang this over your Bed, Cooker, Dog, Front Door

Whatever You Do~ You're Wrong

Cheering Old Saw

LATE NIGHT THOUGHT APROPOS

My word, has not the whole business of masculinity done a graceful pirouette. Nowadays Butch no longer flexes his rippling diaphragm, tears up telephone directories, hits golf balls several hundred yards, double de-clutches and drinks yards of ale without pissing himself – he is the master of the dying feminine arts. Gracious lady, viewing him across a crowded room, breathes huskily 'Now there is a man, he knits, he sews, he cooks Crêpe Suzettes, he changes nappies, he gives excellent suck, he knows his hoover, he shaves his legs, there is a man'. Cry your heart out, Charlie Atlas. You're dead anyway, a fact you keep forgetting.

174

USEFUL PEOPLE AND WHERE TO PIN'EM DOWN

Do not be dismayed if it seems as if it is all happening in London and nowhere else – in most cases the London lot can tell you which is the nearest contact for you outside the great metropolis.

KEEPING FIT
The Keep Fit Association, 70 Brompton Road, London SW3 1HE (01-584 3271).

Weight Watchers, 1 Thames Street, Windsor, Berks SL4 1SW (Windsor [95] 69131).

LADIES
Brook Advisory Centre, 233 Tottenham Court Road, London W1 (01-323 1522)

The Family Planning Association, 27-35 Mortimer Street, London W1A 4QW (01-636 7866).

The National Federation of Clubs for the Divorced and Separated, c/o The General Secretary, 13 High Street, Shelford, Cambridge.

The National Marriage Guidance Council, Little Church Street, Rugby, Warwickshire CV21 3AP (Rugby [0788] 73421).

RIGHTS
Citizens' Rights Office, c/o Child Poverty Action Group, 1 Macklin Street, London WC2 (01-242 6672).

Consumers' Association, 14 Buckingham Street, London WC2N 6DS (01-839 1222).

Legal Action Group Information Service, 28a Highgate Road, London NW5 1NS (01-485 1189).

National Association of Citizens' Advice Bureau (CAB), ·26 Bedford Square, London WC1B 3HU (01-636 4066).

National Federation of Housing Societies, 86 Strand, London WC2R 0EG (01-836 2741).

Office of Fair Trading, Chancery House, 53 Chancery Lane, London WC2 (01-242 2858).

KIDS
Child Minders, 67a Marylebone High Street, London W1 (01-935 2049) – babysitters for London homes.

Families Need Fathers, Dr Alec Elisthorn, 23 Holmes Road, London NW5 (01-485 4226).

Gingerbread, 9 Poland Street, London W1V 3DG (01-734 9014)

National Council for One-Parent Families, 255 Kentish Town Road, London NW5 2LX (01-267 1361).

Pre-School Playgroups Association, Alford House, Aveline Street, London SE11 5DJ (01-582 8871).

HOLIDAYS
Canvas Holidays and Carefree Camping, 7 Wrensfield, Hemel Hempstead, Herts HP1 1RN

Euro Camp Travel Ltd., Tipping Brow House, Mobberly Knutsford, Cheshire (Mobberly |056 587| 2540).

Townsend Thorensen, 127 Regent Street, London W1 (01-734 4431).

Seaspeed Hovercraft Central Reservations, 7 Cambridge Terrace, Dover, Kent (01-606 2894).

CARS
Automobile Association, Fanum House, Basingstoke, Hants RG21 2EA (Basingstoke |0256| 20123).

Royal Automobile Club, P.O. Box 100, RAC House, Lansdowne Road, Croydon, Surrey (01-686 2525).

GENERAL

Alcoholics Anonymous, 11 Redcliffe Gardens, London SW10 (01-352 9669).

Release, 1 Elgin Avenue, London W9, (01-289 1123, night emergency 01-603 8654) for advice on medical, social security, housing and legal problems.

Rentokil Ltd., 16 Dover Street, London W1X 4OJ (01-493 0061).

178

WHERE TO READ MORE ABOUT IT

COOKING
Archard, Merry *Cook For Your Kids* (1975) George Allen & Unwin.

Christian, Glynn *The No-Cook Cookbook* (1974) Jupiter Books.

Patten, Marguerite *How To Cook Book* (1974) Hamlyn Group.

Sewell, Elizabeth *Barbecue Cookbook* (1973) Hamlyn Group.

Tingey, Nancy and Frederick *The Open Air Cook Book* (1975) Letts.

Westland, Pamela *The Everyday Gourmet* (1976) Elm Tree Books.

HOUSEWORK
Bracken, Peg *The I Hate to Housekeep Book* (1963) Arlington Books.

Conran, Shirley *Superwoman* (1975) Sidgwick & Jackson.

Do-It-Yourself Home Encyclopedia (1976) Collins.

Linden, Bruce *Home Maintenance and Improvement* (1973) Queen Anne Press.

…NG FIT
…vis, Adelle *Let's Get Well* (1974), *Let's Eat Right to Keep Fit* (1974) George Allen & Unwin.

The Which? Slimming Guide (1974) Consumers' Association.

LADIES
Comfort, Dr Alex *The Joy of Sex* (1974) Jonathan Cape.

Gregg, Hubert *Come Live With Me* (1974) Bachman & Turner.

Oxford Book of Quotations (1941) Oxford University Press.

Sex With Health (1974) Consumers' Association.

Taylor, John *The Care and Feeding of Young Ladies* (1975) M & J Hobbs/Michael Joseph.

Time Out's Book of London (1973) Time Out Ltd.

RIGHTS AND WRONGS
Claiming on Home, Car and Holiday Insurance (1974) Consumers' Association.

Friedman, Gil *How To Conduct Your Own Divorce In England and Wales* (1975) Wildwood House.

Green, Maureen *Goodbye Father* (1976) Routledge & Keegan Paul.

Krimgoltz, Maxine and Brian J. C. Good *Renting and Letting a Home* (1976) George Godwin Ltd.

The Legal Side of Buying a House (England and Wales) (1965) Consumers' Association.

One Parent Families in Avon, which is published by the British Council for Voluntary Services, 9 Elmerdale Road, Clifton, Bristol BS8 1SW costing 35p.

Ward, Christopher *How To Complain* (1976) Pan Books.

Willmot, Phyllis *The Consumer Guide to British Social Services* (1967) Pelican Books.

KIDS

Activity Holidays in England published by the English Tourist Board, 4 Grosvenor Gardens, London SW1W 0DU.

Brandreth, Gyles *Gyles Brandreth's Complete Book of Home Entertainment* (1974) Shire Publications Ltd.

Harvey, Dr David *The Baby Book* (1975) Marshall Cavendish.

Hinde, Cecilia H. *Floury Fingers* (1962), *Time For Tea* (1973) Faber & Faber.

A useful series of six books: *The KnowHow Book of: Batteries and Magnets, Flying Models, Paper Fun, Print and Paint, Puppets, Spycraft* (1975) Usborne Publishing.

Pelham, David *The Penguin Book of Kites* (1976) Penguin Books.

Practical Camper, a monthly magazine costing 30p.

Vermeer, Jackie and Marian Lariviere *The Little Kid's Four Season's Craft Book* (1975) Nelson.

Whitehorn, Katherine *How to Survive Children* (1975) Eyre Methuen.

CARS

AA Book of the Car (1969) Drive Publications.

Hudson-Evans, Richard *The Drive-In Car Owner's Book* (1976) William Luscombe.

Wickerson, John *The Motorist and The Law* (1975) Oyez Publishing.

GARDENING

The Golden Hands Book of Growing For Cooking (1973) Marshall Cavendish.

Redeaa, Helen and George Seddon *Your Kitchen Garden* (·1975) Mitchell Beazley.

Simons, A. J. *New Vegetable Grower's Handbook* (1975) Penguin Books.

Shewell-Cooper, W. E. *Vegetables – Growing and Cooking The Natural Way* (1975) George Allen & Unwin.

Wright, Michael (Ed.) *The Complete Indoor Gardener* (1974) Pan Books.

GENERAL

Moorehead, Caroline *Helping* (1975) Macdonald & Jane's – a list of useful names and addresses of voluntary organisations which can cope with practically every situation you can think of!

FOR
NOTES
___
OR
OTHER USEFUL ADDRESSES
___
OR
SCRIBBLING PAD
___
OR
TELEPHONE MESSAGES
___
OR
WHATEVER ELSE YOU
USE BLANK PAGES
FOR IN EMERGENCIES
___
OR
SIMPLY A BLANK
PAGE AND HOW TO
FILL IT
___

INDEX

187

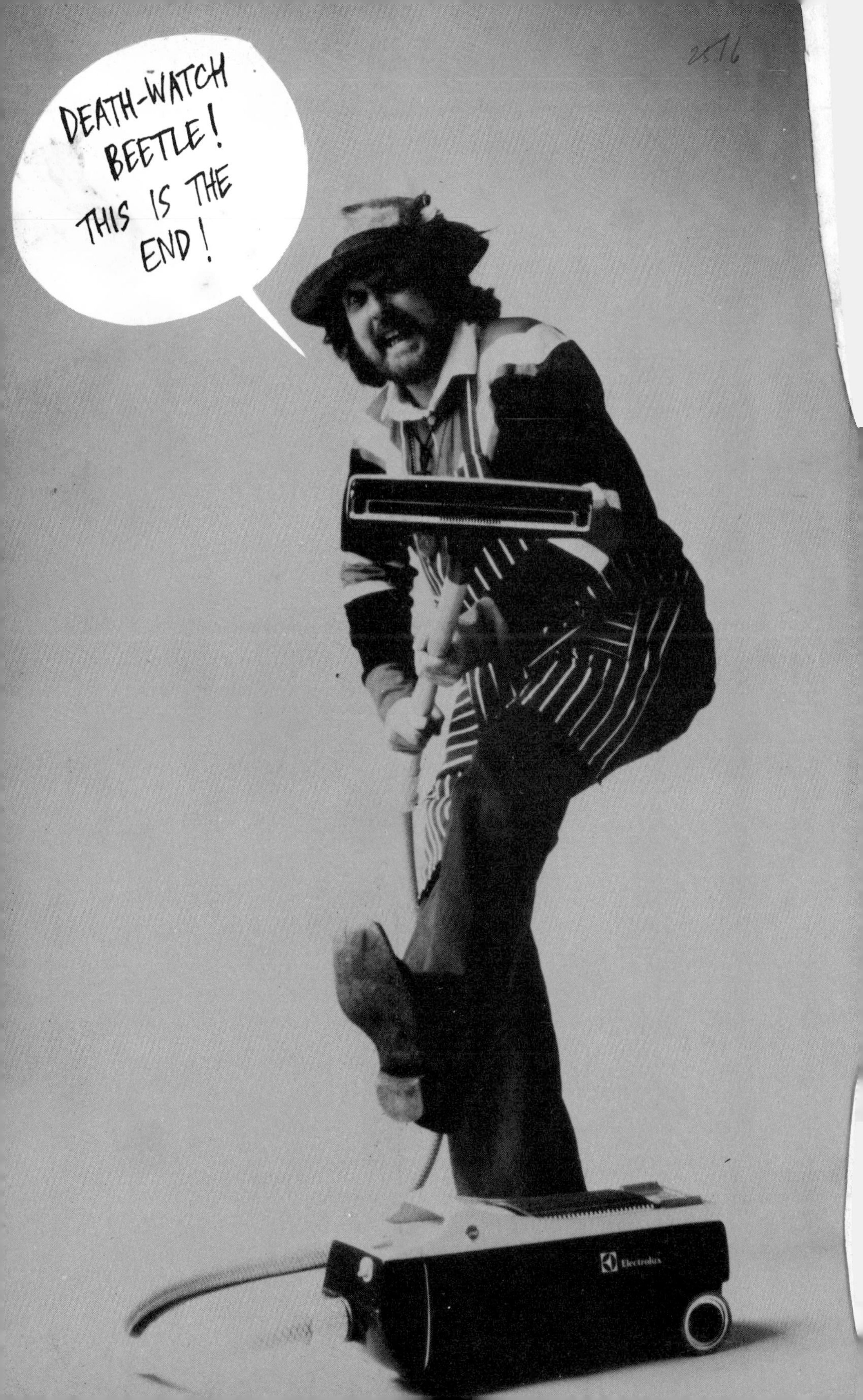
DEATH-WATCH BEETLE! THIS IS THE END!
Electrolux